THE EARTH BROKERS

After decades of failed development plans for the South and the mounting pressure of the environmental crisis all over the planet, the Earth Summit was billed as a dramatic new approach to solving the planet's problems because, for the first time, it was recognized that environment and development were inseparable and thus needed to be tackled together. The recognition of this link, however, turned out to be a double-edged sword, as development quickly became much more important than environment. There was little recognition of the underlying cause of today's crisis – the unsustainable economic models that most of the world is currently following. Free trade, multinational corporations, militarism – some of the biggest contributors to today's crisis – were deliberately left off the agenda. Instead, the Earth Summit attempted to 'green' development and its major promoters by pushing the environment to the top of the agenda. UN and government agencies adopted this new green solution without questioning the assumption that growth and further development were necessary, let alone the assumption that they were possible. Because of this, the Summit was flawed in both conception and execution. As a result, the new order that is emerging after the Rio de Janeiro conference is identical to the old one. If this new order were merely a warmed-over version of the old, things might be expected to continue deteriorating at the current pace, if not accelerate, since the new mantra is that the environment may even be a profitable enterprise that will stimulate development. What is more, the new order is slowly creating a global management élite that is coopting the strongest people's movements, the very movements that brought the crisis to public attention.

Pratap Chatterjee is Global Environmental Editor of the Inter Press Service, Washington, DC. **Matthias Finger** is Associate Professor at Teachers College, Columbia University, New York.

THE EARTH BROKERS

Power, Politics and World Development

PRATAP CHATTERJEE and
MATTHIAS FINGER

Routledge
Taylor & Francis Group

LONDON AND NEW YORK

First published 1994
by Routledge
Published 2013 by Routledge
2 Park Square, Milton Park, Abingdon, Oxon OX14 4RN
711 Third Avenue, New York, NY 10017, USA

Routledge is an imprint of the Taylor & Francis Group, an informa business

Typeset in Perpetua by
Solidus (Bristol) Limited

British Library Cataloguing in Publication Data
A catalogue record for this book is available from the British Library

Library of Congress Cataloging in Publication Data
Chatterjee, Pratap.
The earth brokers : power, politics and world development / Pratap
Chatterjee and Matthias Finger.
p.   cm.
Includes bibliographical references and index.
ISBN 978-0-415-10963-5 (pbk)
1. Environmental policy. 2. Environmental degradation—Government
policy. 3. Environmentalism. I. Finger, Matthias. II. Title.
GE170.C48   1994
363.7'056—dc20   93-41346
                                                            CIP

CONTENTS

CONTENTS

CONTENTS

ACRONYMS AND
ABBREVIATIONS

ABB
Asea Brown Boveri
ACORD
Agency for Coordination and Development
ANEN
African NGOs Environmental Network
ANGOC
Asian NGO Coalition
APPEN
Asia Pacific People's Environmental Network
ARCO
Atlantic Ritchfield Oil
ASCEND21
Agenda of Science for Environment and Development into the 21st Century
BCSD
Business Council for Sustainable Development
Big 10
The major US environmental lobbying organizations
CONGO
Conference of NGOs (in consultative status with the UN Economic
and Social Council)
CNN
Cable Network News
CFC
Chloro-Fluoro-Carbon
CSD
Commission on Sustainable Development
ECOSOC
Economic and Social Council to the General Assembly of the United Nations

EC
European Community
EDF
Environmental Defense Fund
EEB
European Environmental Bureau
ELCI
Environmental Liaison Center International
ENDA
Environment and Development Action in the Third World
FAO
Food and Agriculture Organization
FoE
Friends of the Earth
GATT
General Agreement on Tariffs and Trade
GDP
Gross Domestic Product
GEF
Global Environmental Facility
GM
General Motors
GNP
Gross National Product
G-77
Group of 77
IBRD
International Bank for Reconstruction and Development
ICC
International Chamber of Commerce
ICI
Imperial Chemical Industries
ICSU
International Council of Scientific Unions
ICVA
International Council for Voluntary Agencies
IDA
International Development Agency
IEB
International Environmental Bureau
IFC
International Facilitating Committee
INC
International Negotiating Committee

INGOF
International NGO Forum
IPCC
Intergovernmental Panel on Climate Change
ISMUN
International Youth and Student Movement of the United Nations
ITO
International Trade Organization
IUCN
International Union for the Conservation of Nature (now World Conservation Union)

LDC
Less (or least) developed country

MIGA
Multinational Investment Guarantee Agency

NATO
North Atlantic Treaty Organization
NGO
Non-governmental organization
NRDC
National Resources Defense Council

OECD
Organization for Economic Cooperation and Development

PrepCom (I–IV)
Preparatory committee meetings held prior to UNCED

TFAP
Tropical Forest Action Plan
TNC
Transnational corporation
TQM
Total quality management
TWN
Third World Network

UNCED
United Nations Conference on Environment and Development
UNCSD
United Nations Commission on Sustainable Development
UNCTC
United Nations Center on Transnational Corporations
UNESCO
United Nations Educational, Scientific and Cultural Organization
UNDP
United Nations Development Programme

UNEP
United Nations Environment Programme
USAID
US Agency for International Development
WICE
World Industry Council for the Environment
WICEM I
First World Industry Conference on Environmental Management
WMO
World Meteorological Organization
WRI
World Resources Institute
WWF
Worldwide Fund for Nature

3M
Minnesota Mining & Manufacturing

INTRODUCTION

In New York in December 1989, the member states of the United Nations agreed on Resolution 44/228 – the 228th decision of its Forty-Fourth General Assembly. The Resolution noted with concern that the world's environment was deteriorating rapidly and recommended that the UN General Assembly convene a conference of national leaders of the highest level to save the planet from catastrophe. Officially, this was to be called the United Nations Conference on Environment and Development – UNCED for short. Unofficially it was dubbed the 'Earth Summit' by the man who was chosen to put it together, Maurice Strong, a Canadian businessman and diplomat.

Six months later, the first of four major preparatory committee meetings (the meetings were called PrepComs I to IV) to thrash out conventions and agreements for the leaders to sign at the Summit was held in Nairobi. A member of a non-governmental organization (NGO) attending it sent out a memo by computer to hundreds of other NGOs following the talks describing his own reactions to the name 'Earth Summit'.[1] To him, he said, it conjured up the image of a steep mountain with the heads of state gathered at the summit from where the planet would be saved. The people of the planet were waiting below for the agreements to be signed at the top and brought down to them. In between them and the leaders, bearers toiled, carrying proposals up the mountain.

The next three preparatory meetings were held in Geneva (II and III) and New York (IV). Many NGOs were actively encouraged, and some even financially supported, to attend the meetings. And by the fourth meeting about 1,400 NGOs had officially registered with the UNCED secretariat as observers and lobbyists in the process. Many more followed the negotiations by computer, fax, and regular mail.

After the meetings and the lobbying were finished, the two of us sat down to review what had been achieved over the course of almost two years. This was about two months before the Earth Summit itself was held in Rio de Janeiro in June 1992. We concluded that, as a result of the whole UNCED process, the planet was going to be worse off, not better. We wrote a short paper on the subject and sent it out to hundreds of people to solicit their opinions.[2] Readers wrote in from Massachusetts and Michigan to Mongolia, and others translated our paper into French, Spanish, and Swedish. Almost all agreed with our critical assessment, but said that they had not seen anyone else actually put such a strong thesis on paper. We decided that we needed to set out our thoughts in much more detail for other people who did not have the opportunity to participate in the two-year process that led leaders of governments, industry and NGOs to Rio, but actually failed to take them to the summit of the mountain from where to save the planet. This book is the result.

In it we offer a comprehensive and critical overview of the entire UNCED process. We look at its origin, its context, and the major agents involved, as well as its outcomes. But because UNCED is at the core of the recent developments in the environment and development arena, this book actually reaches beyond UNCED. And because UNCED occurred at a crucial moment in environmental and developmental history, this book also helps readers understand the transformation of 'development' and the recent quite profound changes in North–South relations, as well as the deep changes the Green movement has undergone.

In the first part we highlight the context and the process of UNCED. We present and critically analyze the main documents that have been written in preparation to that process, as well as the ones that have come out of it. Parts II and III look at the main non-state players in the UNCED process, i.e. non-governmental organizations on the one hand, and business and industry on the other. Indeed, traditionally everybody has been looking at governments as being the major agents. However, as we hope to show, governments are only part of the picture: the corporate sector and some NGOs have come to be equally important agents in the UNCED process. Part IV looks at the financial and institutional outcomes of the UNCED process, and assesses what, on that basis, we can expect for the future. Finally, we conclude with an analysis of what that means for the planet.

Throughout this book we show how UNCED has promoted business and industry, rehabilitated nation-states as relevant agents, and eroded the Green movement. We argue that UNCED has boosted precisely the type of industrial development that is destructive for the environment, the planet, and its inhabitants. We see how, as a result of UNCED, the rich will get richer, the poor poorer, while more and more of the planet is destroyed in the process.

THE DEVELOPMENT PARADIGM

In order to understand the UNCED process, it must be located in the larger context of the development paradigm and it therefore must be looked at from a broader historical perspective. Most important of all, UNCED must be seen in the context of industrial development, a process that can be traced back to the Industrial Revolution and beyond. Indeed, the idea of development is rooted in the Enlightenment ideal of a rational society of free and responsible citizens, i.e. ultimately a society governed by scientific principles and managed accordingly. The emergence of industrial production in the nineteenth century was rapidly incorporated into this development paradigm: industrial development came to be seen as a means — so to speak the motor — of making this modern and rational society come true. Unfortunately, the means turned into an end, development became a goal in itself. This is what we call the development ideology or paradigm.

Marxists have criticized industrial development since its social effects started to be felt in the late nineteenth century. They criticized it on the grounds that it produces injustices, enhances unequal power structures and exploits people. However, Marxists have never questioned the underlying idea that industrial development will free society from the constraints of nature, and thus ultimately liberate people altogether. The main obstacle that prevented this process from happening was not to be found in the development process itself, Marxists argued, but rather in the political power structures, which were perpetuating inequities and oppression. Marxists, therefore, remain caught in the development paradigm.

After the experience of the First World War, and even more so after the Second World War, isolated individuals expressed their doubts as to whether there was not something fundamentally wrong with this industrial development process:

were the two wars simply accidents of history or were industrialization and modernization leading to precisely the type of barbarism seen in the two conflicts? Marxists of a new kind – the so-called Critical Theorists – rapidly took the upper hand in voicing these doubts. Though they questioned whether the declared emancipation of humankind, promised since the Enlightenment, was ever going to be realized, they attributed this failure to politics, rather than to the development paradigm. Basically they thought that advanced capitalist societies were developing particularly vicious and hidden ways to oppress men and women. As a result, humanity would miss the unique opportunity to liberate itself that industrial development offered. Thus, even after the Second World War techno-scientific industrial development remained an unquestioned tool even for the most vocal critics of modern society.

What is more, in an effort of collective denial promoted by a massive public relations campaign, further industrial development was declared, in the aftermath of the Second World War, to be also the means of bringing about peace among nations. As a result, the United Nations was set up with the mission to promote 'peace through development'. No longer was industrial development simply going to lead to a modern and rational society, it was also going to bring peace to the world. With the United Nations promoting it, industrial development progressed exponentially and planet-wide. What is more, the aggressive reconstruction of Western Europe became the model for the industrialization of the entire world. Development was now clearly the goal, and the development process of the North, spearheaded by the USA, was to be replicated by the South. The rare humanists who feared that the human side would get lost in the process were silenced, as the 'cultural subsystem' was singled out and declared to be the realm of truly human aspirations. Thus, culture became a luxury that was made possible by continuous industrial development.

THE COLD WAR

The Cold War is the next important element to consider in order to understand the process of industrial development. First, the Cold War became one of the driving forces of industrial development, because it stimulated scientific and technological progress on the one hand, and promoted military-induced industrial production on the other. Second, the Cold War cemented the nation-state system and thus reinforced the idea that nation-states were the most relevant units within which problems had to be addressed. Therefore, the nation-states were also seen as the primary agents of development, the 'development agencies', so to speak.

Indeed, because of the Cold War, the nation-states continued to be seen as the units within which development occurs and must be promoted, because it is economic and military strength that defines each nation's relative power. In promoting the Cold War, nation-states remained the key agents for at least another forty years. Again, industrial development came to be seen as a means to enhance national power, thus hiding the fact that the means had overtaken the ends.

THIRD WORLD DEVELOPMENT

The development paradigm was further strengthened by the political independence of many Third World countries. Indeed:

Truman [had] launched the idea of development in order to provide a comforting vision of a world order where the US would naturally rank first. The rising influence of the Soviet Union – the first country which had industrialized outside capitalism – forced him to come up with a vision that would engage the loyalty of the decolonizing countries in order to sustain his struggle against communism. For over 40 years, development has been a competition between political systems.[3]

With the Cold War solidly established and entirely embedded in the post-war reconstruction and the Third World build-up, the development paradigm became institutionalized in the very structure and nature of Third World nation-states. Thus these countries started to enter the industrial circuit by borrowing money and exporting raw materials. Given Third World independence and the context of the Cold War, the nature of industrial development was not questioned until the late 1960s. Only then did social movement activism begin to raise serious doubts as to whether industrial development would really lead to the type of society promised by Truman and others.

THE SOCIAL MOVEMENTS OF THE 1960s

In the North, the social movements of the late 1960s emerged within the context of already high levels of industrial development. The main critique they voiced was the oppressive and technocratic tendencies of development, i.e. the danger that the people, the human side, would get lost and forgotten. One must distinguish between the American version of social movement activism and the European one. If the American version is a product of the counter-cultural movement, the European movement is a product of the New Left. Both agree that the process of development has got out of human control and does not serve the majority of the people. The counter-cultural movement formulates a cultural critique: it is concerned with the values brought forth by the development process and seeks to substitute these with more human values. The critique formulated by the New Left, in contrast, is in essence political. It is a critique of oppression, domination, and exploitation. Consequently, more participation, more democracy, and more involvement of the citizens in decision-making are seen by the New Left as the answers to the shortcomings of industrial development. During the late 1960s, however, neither the counter-cultural movement nor the New Left questioned the process of industrial development, though both were unhappy with its inhuman consequences.

The political critique formulated by the New Left in the North is actually quite similar to the critique voiced in the South, where social movements were also calling for a more participatory form of development. Development, in the South, attracted criticism in the late 1960s and the 1970s on the grounds that it was top-down, exploitative, and oppressive. The national and local élites in the South were mainly seen as the longer arm of the North, of Northern governments, and of Northern multinationals. Opposing this, the social movements in the South were advocating 'another', i.e. a more participatory, more human-centred, and more indigenous form of development. Some went as far as to suggest breaking links with the North and promoting self-reliance. However, for all the radical critiques of Northern-centredness and Northern-drivenness, development was being questioned in the South by only a very few people in the late 1960s and throughout the 1970s. It was not until the advent of the Green movement in the North, in the 1970s, that a new argument was added to the critique of industrial development.

THE GREEN CRITIQUE OF DEVELOPMENT

Before the early 1970s it is difficult to identify a coherent Green critique of development. Of course, since the end of the nineteenth century there have been various nature protection organizations. Since the 1930s some scientists and engineers have focused on natural resources conservation and environmental management, starting with forestry and specific ecosystems. After the Second World War two big international organizations were created along such conservationist ideals – the International Union for the Conservation of Nature, now called World Conservation Union (still referred to as IUCN), and the World Wildlife Fund, now called Worldwide Fund for Nature (WWF). Within the UN system the environment was equated with science and attributed to UNESCO, the United Nations Educational, Scientific and Cultural Organization. One can say that until the late 1960s (scientific) environmentalists hardly questioned development. Rather, they were concerned with species conservation and rational resources management in line with the overall development paradigm.

But in the early 1970s, in the context of the social movements, one can detect, in the North, the replacement of conservationist ecology with political ecology. It was under the influence of the New Left, in particular, that environmental problems become politicized and prominent. In addition to natural resources issues, this politicization focused primarily on pollution problems such as oil spills, chemical hazards, and nuclear pollution. In 1972 the Club of Rome, a group of concerned leaders from business, academia and government, published its *Limits to Growth*, highlighting in particular the possible input limits to further industrial development. In the same year the UN held its first Conference on the Human Environment, in Stockholm. Again, the focus was on natural resources management and, to a lesser extent, on pollution control, as both resources depletion and pollution were seen as potentially jeopardizing development.

Within the intellectual context of the New Left, environmental problems remained mainly political problems. Resources depletion and pollution were, it was argued in the 1970s, the result of existing power structures, which oppressed nature and people alike. Because of this political framework, political ecologists remained uncritical of many of the destructive forces of industrial development, in particular of modern science, high technology, and the nation-state. Indeed, their markedly Northern-centred view led political ecologists to propose scientific progress, better technologies, and especially better policies as the answers to resources depletion and pollution problems. The nation-state remained, in their view, the most important, if not the only, relevant unit of action.

It was at this time, within the context of political ecology, that most environmental agents emerged. Be it Greenpeace, Friends of the Earth (FoE), the National Resources Defense Council (NRDC) in the USA, or many more, they all refer to this framework of political ecology within which they operate and which they perpetuate. Later in the 1970s Green parties used this Green movement in Western Europe while simultaneously strengthening the purely political approach to environmental issues and problems. Therefore, because of the political ecology framework, the nation-state remained the focus of environmental activists. The causes of environmental degradation were thus localized in politics and not, for example, in the dynamics of the industrial development process. Yet this analysis not only ignored the root causes of the development crisis, it also suggested that further scientific, technological, social, and political development would help solve the problems. In short, though it added some arguments to the critique of development, the Green movement of the 1970s did not identify industrial development as being the problem for the planet and its inhabitants.

THE NEW COLD WAR AND GLOBAL ECOLOGY

With the emergence of the New Cold War in the late 1970s, fear and anxiety about a possible nuclear holocaust overshadowed environmental concerns in the North. But interestingly, the New Cold War prepared the ground for global ecology, for which the so-called theory of the nuclear winter was probably a trigger. First put forward in 1982, this theory states that a nuclear explosion anywhere on this planet has the potential to induce climate change planet-wide. Rather than being about the nuclear threat, this theory is in fact about global environmental change. As such it was symptomatic of a whole new approach to environmental problems emerging at the beginning of the 1980s: global ecology.

Ozone depletion and global warming, in particular, along with other global environmental issues such as deforestation and soil erosion, became the focus of this new global ecology. Global ecological problems were no longer simply resources depletion or pollution issues. Indeed, in addition to pollution problems and input limits to growth, global ecology now also pointed to potential global output problems of industrial development. It now appeared that such output limits might actually be far more serious than the input limits and the pollution problems, for which there are, to some extent, technological and political solutions. We think that the global ecology of the early 1980s was actually a far more serious challenge and critique of industrial development than anything else that came before.

THE CHALLENGES OF GLOBAL ECOLOGY

The real effects of global ecology only became apparent when the Cold War ended in 1986 with Gorbachev coming to power in the Soviet Union. It was at this time that the possible consequences of global ecology really hit home: global ecology questions the very essence of industrial development, and therefore also the agents that live off this process. Among the first major agents to be challenged are of course business and industry, especially big business such as multinational corporations. Indeed, if the challenge of global ecology is taken seriously, there are now serious output limits to further economic growth and industrial development. Promoting such massive industrial development, as most of these multinational companies do, amounts to promoting accelerated destruction of the global environment.

A second type of agent whose pursuit of industrial development is being challenged are nation-states. Protected by the Cold War and legitimized by the social and environmental movements of the late 1960s and the 1970s, in the age of global ecology nation-states not only have a legitimation problem, they also now have to demonstrate that they are still relevant agents when it comes to the new challenges of global ecology. Are they indeed able to address the challenges raised by global ecology successfully?

The role of the military is of course brought into question in a very similar way by global ecology. Indeed, in the light of the new global environmental changes and challenges, issues of national security increasingly seem to be irrelevant. As a result, the military-industrial complexes of the world are now figuring out ways and means to make sense of themselves in the eyes of an increasingly critical public.

And if industrial development, business and industry, nation-state structures, national governments, and military-industrial complexes are increasingly brought into question by global ecology, Southern élites are hardly better off, as they basically derive their power and privileges from imitating the North and its industrial development model. If further industrial development is made impossible by global change and challenges, Southern élites are threatened. A similar threat extends to the UN system whose aim, as we have seen, is to promote development – not to mention the fact that over the past forty years the UN system has created a development élite of its own, whose very existence is now brought into question by the global ecological threat.

And finally, the Green movement, too, is brought into question by global ecology and its challenges. Having its roots in either conservation or political ecology, the Green movement needs to redefine itself, as it is no longer obvious that the traditional problem-solving approaches it promoted are still valid when applied to the new global environmental challenges. Moreover, the Green movement also has to find a new acceptance in the eyes of a concerned public as a relevant agent in this new global environmental arena.

In this book, we show that UNCED offered a unique opportunity to all these different agents to redefine and relegitimize themselves in the new age of global ecological changes and challenges. Some have done better than others. But overall, as we argue, the outcome is not a better way to address the global ecological crisis. Rather, the outcome is a new push for more environmentally destructive industrial development.

Two publications have become particularly important, as they try to reassess some of these agents' roles in the light of the new challenges. Both are the products of international commissions: the so-called 'Brundtland report' entitled *Our Common Future* is the outcome of the World Commission on Environment and Development, created by the UN in 1983, while the report entitled *The Challenge to the South* is the product of the South Commission, established in 1987. Both were written in time for the Rio conference. Let us look at them first.

Part I

THE DOCUMENTS

1

WHOSE COMMON FUTURE?

The essence of the philosophy of the World Commission on Environment and Development can actually be found on the very first page of the Brundtland report. This report, the Commission says:

is not a prediction of ever increasing environmental decay, poverty, and hardship in an ever more polluted world among ever decreasing resources. We see instead the possibility for a new era of economic growth, one that must be based on policies that sustain and expand our environmental resource base. And we believe such growth to be absolutely essential to relieve the great poverty that is deepening in much of the developing world. . . . We have the power to reconcile human affairs with natural laws and to thrive in the process. In this, our cultural and spiritual heritages can reinforce our economic interests and survival imperatives. . . . This new reality, from which there is no escape, must be recognized and managed.[1]

We cannot in this book go into the history of how the UN system created the World Commission on Environment and Development. Nevertheless, let us briefly recall here the context within which the Brundtland Commission emerged. It is the context of the New Cold War and the re-emerging East–West conflict at the beginning of the 1980s. It is against this threat to 'our common security' – highlighted by the debate about the Euromissiles, as well as by the nuclear winter theory – that the Brundtland Commission was created. Not surprisingly, the title of the Brundtland report, *Our Common Future*, is very similar to the title of the Palme report, *Our Common Security*, whose main concern was the nuclear threat.[2] As a matter of fact, the Brundtland report devotes an entire chapter to a quite radical critique of the arms race, to conclude that 'the nations must turn away from the destructive logic of an "arms culture" and focus instead on their common future'.[3] We also note that the Brundtland report actually remains the only document in the entire UNCED process that explicitly deals with the military as a problem. This can be explained by the fact that the Brundtland Commission was born in the context of the Cold War.

What is more, the Brundtland Commission sees at least part of its role as helping to break out of the international deadlock caused by the Cold War. In her preface, ex-Premier Brundtland says: 'After a decade and a half of standstill or even deterioration in global cooperation, I believe the time has come for higher expectations, for common goals pursued together, for an increased political will to address our common future'.[4] It might well be that in the initial phase of the Commission the environment was actually more of a rallying point to foster cooperation among nation-states than the real common challenge.

In the process of its work, the Commission identified the real challenges as population and human resources, food security, species and ecosystems, energy, industry, and the urban challenge. But by breaking down the environmental question into these six challenges, the Brundtland Commission managed to redefine the global environmental crisis in terms of a problem that can be solved by nation-states and their cooperation in promoting economic growth. And such growth, the Commission says, can essentially be achieved by manipulating and improving technology and social organization.[5] Overall, one can say that not much thinking seems to have gone into the analysis of the real causes of today's crisis. The major concern does not seem to be the crisis, but the potential conflicts between nation-states that could arise because of a lack of development. Let us now look at each of the six challenges the Commission has identified in more detail.

POPULATION

In the beginning of its section on population, the Brundtland report states that 'present rates of population growth cannot continue'.[6] And: 'Nor are population growth rates the challenge solely of those nations with high rates of increase. An additional person in an industrial country consumes far more and places far greater pressure on natural resources than an additional person in the Third World'.[7] Despite these statements, the analysis put forth by the Commission on population issues is, in our opinion, basically flawed. It rests on the assumption of two fundamental relationships, both of which must be balanced: there should be a balance between population size and available resources on the one hand, and between population growth and economic growth on the other.[8]

Population is basically seen as an input problem at the national level. The question is whether there are enough natural resources to sustain a certain number of people within given national boundaries. There is also mention, in the report, that people should have equitable access to the overall resources pool, as such equitable access as well as further economic growth are both important means to get fertility rates down. Says the Commission: 'sustainable economic growth and equitable access to resources are two of the more certain routes towards lower fertility rates'.[9] In other words, lowering fertility is seen by the Commission as being achievable through social and economic development alone.

Since 'almost any activity that increases well-being and security lessens people's desires to have more children than they and national ecosystems can support',[10] the second strategy envisaged by the Commission is to balance population growth rates with economic growth rates. Starting with the realistic assumption that populations will continue to grow, the Commission advocates higher economic growth as well as better education – called 'improving the human potential' – and technological improvements in order to make more efficient use of the available natural resource base, or even enhancing this natural resources base. Again, this is achievable through economic growth. Overall, then, the Commission's main recommendation for dealing with population growth is more development: 'A concern for population growth must therefore be part of a broader concern for a more rapid rate of economic and social development in the developing countries.'[11]

While the Commission certainly pursues the laudatory aim of providing equitable access to resources, this is combined with advocating further growth in order to raise the poor to the levels of the rich. Yet, this is a dangerous idea because the Commission's own figures show that the rich are consuming the vast bulk of resources, which is the major reason for the present crisis to begin with. The Commission's own figures show, for example, that the populations of the Northern countries, with a quarter of the world's inhabitants, consume fifteen times as much paper as their counterparts in the South. Demand from the poor for fuelwood is another major cause of deforestation, but given that the numbers of people consuming trees for paper, furniture and construction purposes are much smaller than those felling them for fuel, and the proportion of wood used is considerably higher, it would surely be more effective to act in the North first.

None of these ratios is at all new. The economist E.F. Schumacher used similar figures in his famous book *Small is Beautiful*, published in 1973.[12] He showed that the United States with 5.6 per cent of the world's population was consuming 63 per cent of the world's natural gas, 44 per cent of the world's coal, 42 per cent of the world's aluminium, and 33 per cent of the world's copper and petroleum, all non-renewable resources. He said:

It is obvious that the world cannot afford the USA. Nor can it afford Western Europe or Japan. In fact, we might come to the conclusion that the Earth cannot afford the 'modern world'. . . . The Earth cannot afford, say, 15 per cent of its inhabitants – the rich who are using all the marvellous achievements of science and technology – to indulge in a crude, materialistic way of life which ravages the Earth. The poor don't do too much damage; the modest people don't do much damage. Virtually all the damage is done by, say 15 per cent . . . The problem passengers on spaceship Earth are the first class passengers and no one else.

In the Brundtland report and in many other reports similar ratios can be found for the consumption of most resources and for the production of most pollutants. But, after quoting such figures, the Commission fails to draw the logical conclusions. It even misses the real point, since it concludes that poverty is the cause of environmental degradation and that higher living-standards will therefore reduce population growth and wasteful consumption. The Commission clearly does not seem to understand that economic growth leads to more consumption and that more consumption leads to more pollution. Even the currently accepted indicators of national income show that those activities that lead to the quickest economic growth cause an increase in pollution. For example, the World Bank reports that 'environmentally benign activities usually contribute a smaller portion to national income than do environmentally malignant ones'.[13] Had the Commission realized this and not been blinded by the development myth, it might have concluded that redistribution and de-industrialization would serve the global environment better than further economic growth.

FOOD SECURITY

Under this heading the Brundtland Commission expresses its concern about how to feed the planet's growing population. The report goes through a wide variety of statistics to show that most of the world has too little to eat despite the fact that food production has continuously outstripped population growth. It also discusses a series of environmental problems impacting negatively on global food production, such as soil erosion, soil acidification, deforestation, and desertification, as well as soil and water pollution. Yet, very optimistically, the report states that 'global agriculture has the potential to grow enough food for all'.[14] Let us see how the Commission comes to such a conclusion and how it conceives of global food security.

Given the Commission's assumption that there is enough food, it sees food security basically as a distribution problem. And such a problem can, of course, be solved by better management, especially on the 'ultimate scale of distribution', i.e. the global scale. In addition, food security is also seen as a traditional political problem, especially on the level of national agricultural policy. The argument of the Commission is in fact very close to the argument we can see in GATT: it is specially subsidized production which is seen as being environmentally (and economically) damaging, since subsidies (in the North) lead to surpluses, which depress international market prices, which in turn 'keeps down prices received by Third World farmers and reduces incentives to improve domestic food production'.[15] In short, it states that it is 'the short-sighted policies that are leading to degradation of the agricultural resource base'.[16] There is no mention of the skewed system of food production such as monocultures, the loss of seed varieties, multinational control, land owner-ship, and much more.

Cursory mention is made in the report of the fact that most of the planet's scientifically stored genetic material is in the hands of Northern laboratories and that private companies are increasingly seeking proprietary rights to improved seed varieties while ignoring the rights of the country they were imported from. Only a few years ago, India for example still had some 30,000 varieties of rice, all of which had different functions and were adapted to different climatic and other conditions. Today, only fifteen varieties cover three-quarters of the country.[17] If the native crops are slowly destroyed or forgotten, and the world's poor have to depend on expensive, less robust and imported seeds, they will never be able to support themselves.

Overall, the problem of Southern agricultural exports is badly fudged in the Brundtland report. While the Commission spends a fair amount of time on the subject of the North dumping subsidized grain in the South, there is hardly any correlation drawn between hunger and poverty and the fact that large private land holdings in the South are being used to grow cash crops for export to the North, rather than feeding the people in the country. The one section on the subject points out that during the 1983–84 famine in the Sahel, Burkina Faso, Chad, Mali, and Niger harvested record amounts of cotton, i.e. 154 million tons of cotton fibre, a sevenfold increase over the harvest in 1962. At the same time, the Sahel region set a record for cereal imports, i.e. 1.77 million tons, up almost nine times over a corresponding period of just over 20 years. The Commission does not draw a conclusion from this, nor does it mention that, simultaneously, world cotton prices have been steadily falling.

The answer of the Brundtland Commission is to emphasize economic growth, export diversification, commodity agreements, and other subsidy policies so that people can actually afford food. As Brundtland points out, the Southern countries cannot compete against Northern food exports because their prices are artificially lowered by subsidies like the EC's Common Agricultural Policy. But countries have to realize that they face a Catch 22 situation. They can only buy this cheap food with foreign exchange, which they can only get by selling cash crops and natural resources at steadily falling prices, thus accelerating the erosion of local self-sufficiency. Yet, would it not be better to take a lesson from the decade-long nosedive in prices and stop depending on exports? Why should a country spend its valuable foreign exchange buying food and selling cash crops whose prices are falling? Does it not make sense to grow the food for the local people first?

In short, the Commission regards the problems as basically technical and political ones, such as the poor design of irrigation systems, the incorrect application of agricultural devices, subsidy allocation, and so on. The problem, however, is systemic. The report, moreover, takes population and its growth as given. The challenge is not, as Brundtland suggests, 'to increase food production to keep pace with demand'.[18] In doing so, the Commission basically imagines a technofix: 'new technologies (will) provide opportunities for increasing productivity while reducing pressures on resources'.[19] To sum up, the Commission envisages some sort of second Green Revolution, which, this time around, will not only be managed globally, but moreover include local people, especially women, in the overall management scheme. To recall, the Green Revolution subsidized the buying of seed, fertilizers, and pesticides, but of course the only people who could afford to buy these were the ones who had access to capital and were then rewarded with large profits. The poor ones who bought into this scheme were poorer as a result of it. In many countries the Green Revolution failed completely because, in addition, the new crops were totally unsuited to the land and caused further famine.

SPECIES AND ECOSYSTEMS

The way the Brundtland Commission talks about species and ecosystems is actually symptomatic of the way it sees biological diversity, nature, and the biosphere: nature is basically viewed as an economic resource to be used for further development. Again, there is a big discrepancy between the diagnosis and the proposed solutions.

The Commission recognizes the alarming rate of species extinction, which is 'hundreds of times higher and could easily be thousands of times higher than the average background rate of extinction'.[20] But having diagnosed this, it immediately downplays the issue – 'extinction has been a fact of life since it first emerged'[21] – and offers a highly unsophisticated analysis of the causes of species extinction and, by extension, of environmental degradation. All damage to the environment, it says, is caused by so-called 'human activities'. The most sophisticated the Commission gets in identifying the causes of environmental destruction is when it blames 'large populations', 'poverty', and 'shifting agriculture'. It also mentions the role of logging policies of many countries that encourage timber exports and livestock ranching.

As a consequence of this very weak analysis, the proposed actions necessarily remain quite general and ideological. For the Commission, the 'first priority is to establish the problem of disappearing species and threatened ecosystems on political agendas as a major economic resource [sic!] issue'.[22] In other words, the priority is to reframe environmental destruction in terms of national economic development policies. Thus, plants, animals, micro-organisms, and the non-living elements of the environment on which they depend become 'living natural resources', which are, moreover, 'crucial for development'.[23] Tropical forests, for example, become 'reservoirs of biological diversity' waiting to be 'developed economically'.[24] In short, the answer the Commission proposes in response to species extinction and habitat destruction, for example, is basically to put species, biodiversity, and nature overall on to the national and international development agenda, i.e. to make them resources for development.

Consequently, species should be managed like all other natural resources, possibly by making use of new technologies, such as bioengineering. The Commission even goes as far as to propose a 'gene revolution' to succeed the Green Revolution, which, as we have seen, was a disaster. Heavily influenced by conservationist environmentalists, in particular WWF and IUCN, the Commission proposes more parks and wildlife conservation areas as the answer. However, in contrast to the 1950s and the 1960s, when species were 'parked' in such areas, in today's new approach species protection must be linked to development. Says the Brundtland report: '. . . governments could think of "parks for development" [sic!], insofar as parks serve the dual purpose of protecting for species habitat and development processes at the same time'.[25] To be sure, such species protection and development is, before all, a national task.

It is clear that the Commission does not adequately analyse the causes of species extinction in particular, and environmental degradation in general. Therefore, many of the solutions the Commission proposes are, in our view, still causes. For example, the Commission applauds the efforts of the World Bank and the US Agency for International Development (USAID) in paying for conservation, but curiously fails to mention that these are two major subsidizers of timber and agricultural export as well as resettlement policies, both leading to species extinction.

The main focus is on national and international management. Community knowledge is basically ignored. Rather, the public needs to be educated, it says, but it fails to notice that the public may once have known all of this or may still know some of it. It suggests, instead, that these people should be required to learn intensive agricultural methods using more fertilizers and pesticides, ignoring the recommendations of the previous chapter on food security, which pointed out that these chemicals were contributors to species extinction.

Overall, the Commission seems to ignore that there is such a thing as an ecological rationality. It has no sense that this might contradict the economic rationality which the Commission imposes upon everything, be it species, biological diversity, ecosystems, or nature.

SUSTAINABLE INDUSTRIAL DEVELOPMENT

The Commission seems to be perfectly aware of some negative environmental consequences of industrial development, and related energy production. It mentions in particular hazardous waste, chemical and nuclear risks, soil, air and water pollution, as well as climatic change. On the other hand, the Commission never mentions negative social and cultural consequences of industrial development. Of course, for the Commission industrial develop-ment is not only desirable, it is imperative. And industry is the key: 'Industry is central to the economies of modern societies and an indispensable motor of growth. . . . Many essential human needs can be met only [sic!] through goods and services provided by industry. The production of food requires increasing amounts of agrochemicals and machinery'.[26] In other words, there are growing needs, and the growth of industry, so the argument goes, is the only way to satisfy these needs. Sustainable development therefore means, in essence, sustainable industrial development. An annual 3 per cent global per capita GDP growth is 'regarded in this report as a minimum for reasonable develop-ment'.[27]

Everything the Commission writes about – in this case waste reduction, pollution control, risk management, and energy consumption and efficiency – must be seen against the background of this industrial growth imperative. All these measures should at least not cut into growth, but if at all possible enhance growth. The main question for the Brundtland Commission, therefore, is how to sustain industrial development without cutting into the resources upon which future growth depends. This is the definition of sustainable develop-ment: 'sustainable development is development that meets the needs of the present without compromising the ability of future generations to meet their own needs'.[28] The main way to achieve this, according to the Commission, is therefore increased efficiency resulting from technological improvements. 'The Commission believes that energy efficiency should be the cutting edge of national energy policies for sustainable development'.[29] But even this efficiency argument must be seen against the background of the growth imperative: the argument says, in essence, that the same or more economic and industrial growth can be achieved with less energy (and material input). The goal therefore is growth and not an ecologically sustainable level of industrial production.

If in relative terms the energy input per capita GNP increase diminishes, absolute consumption of energy therefore will still grow. The following quote illustrates this argument: 'The woman who cooks in an earthern pot over an open fire uses perhaps eight times more energy than an affluent neighbour with a gas stove and aluminum pans. The poor who light their homes with a wick dipped in a jar of kerosene get one fiftieth of the illumination of a 100-watt electric bulb, but use just as much energy'.[30] However, this whole efficiency argument developed against the background of the growth imperative is basically flawed: of course a 100-watt light bulb is 50 times more efficient than a wick dipped in kerosene. And of course a gas stove is about eight times more efficient than cooking over an open fire. However, the argument does not take into account all the energy that was needed to build and is still needed to maintain the entire natural gas and electric infrastructure to begin with. Not to mention the fact that the efficiency argument only refers to technological improvements, neglecting social and cultural consequences of such industrial development.

In short, the efficiency argument developed by the Brundtland Commission – be it technological, economic, or organizational efficiency – only makes sense against the background of sustained industrial development. It is indeed questionable whether at a given level of industrial development substantial resource and energy saving technological improvements can actually be made, and whether in a pre-industrial society, for example, cooking over a woodstove is not the 'most efficient technology'. In any case, the Commission considers that technological improvements leading to more efficiency can only be made by further industrial development, and not by looking at past experiences.

FROM MILITARY TO ENVIRONMENTAL SECURITY

Of all the agents involved in the UNCED process, the Brundtland Commission is the only one to have explicitly addressed the military. This can be explained by the fact that the Commission took as a reference point the Brandt report on North–South relations and the Palme report on the nuclear predicament. As a matter of fact, there is no doubt that the Brundtland Commission and its mandate are heavily conditioned by the overall East–West context of the early 1980s, i.e. by the so-called Euromissile crisis. The Palme report discussed the resulting threat in *Our Common Security*,[31] and the Brundtland Commission was actually much influenced by the same threat, as the title *Our Common Future* suggests. It was against this threat – perhaps best exemplified in the theory of the 'nuclear winter', which also appeared for the first time in 1982 – that the Brundtland Commission emerged. Since then the global environmental questions have remained focused on this nuclear threat. As the Brundtland Commission says, its work occurred against the background of a 'widespread feeling of frustration and inadequacy in the international community about our own ability to address the vital global issues and deal effectively with them'.[32] Indeed, the Commission was above all concerned that environmental degradation could become an additional source of political conflicts. Says the Commission:

Nations must turn away from the destructive logic of an 'arms culture' and focus instead on their common future. The level of armaments and the destruction they could bring about bear no relation to the political conflict that triggered the arms competition in the first place. Nations must not become prisoners of their own arms race. They must face the common danger inherent in the weapons of the nuclear age. They must face the common challenge of providing for sustainable development and act in concert to remove the growing environmental sources of conflict.[33]

We fully agree with the Commission in its opinion that war and security problems have created major environmental stress by, for example, displacing people from their homes. We note with equal discomfort the fact that military spending equals the income of the poorest half of humanity and that more than half of the world's scientists are engaged in research for this. We laud the fact that the Commission has pointed a finger at the 'military-industrial complex' and at the fact that military expenditure is more 'import-intensive' and creates few jobs.

But we also note that the 'arms culture' as the Brundtland Commission calls it is not analysed. The financiers and profit-makers of the arms race are not mentioned. Given that the military is one of the largest polluters in the world in the amount of toxic waste it produces, the energy it consumes and the pain and death that its products cause, surely the Commission should have also talked about the importance of regulating the military industries. Instead, it hands the issue over to international agreements and cooperation to create more security and reduce the need for weapons, certainly a must, but much less effective than committing national governments to stop encouraging the production, the import, and the export of weapons.

As a result of this lack of analysis of the military-industrial complex and its role in industrial development, the chapter of the Brundtland report on peace and security turns into a way of redefining environmental problems in security terms. By considering environmental degradation as yet another cause of conflict among nation-states – which is the basic political unit the Commission considers – the concept of security is enlarged and applied to the environment as well. Says the Commission: 'Action to reduce environmental threats to security requires a redefinition of priorities, nationally and globally. Such a redefinition could evolve through the widespread acceptance of broader forms of security assessment and embrace military, political, environmental, and other sources of conflict'.[34]

Political and environmental sources of conflict are therefore put on the same level and made comparable which, of course, they are not. But by considering the environment as a security issue along with other political issues, the causes of such environmental conflict can be acted upon, it is argued, in the same way as the causes of political conflict, i.e. among others through more development. As a result, military spending is weighed against spending for development. Says the Commission:

The true cost of the arms race is the loss of what could have been produced instead with scarce capital, labor skills, and raw materials. . . . Nations are seeking a new era of economic growth. The level of spending on arms diminishes the prospects of such an era – especially one that emphasizes the more efficient use of raw materials, energy, and skilled human resources.[35]

In short, in this analysis the military is simply an impediment to future development. It is basically for this reason, not for its environmentally and culturally destructive consequences, that the Brundtland Commission criticizes the military. Common environmental security therefore becomes an issue of redirecting the money from the military to development, especially sustainable development. Striving for common security therefore becomes identical to striving for sustainable development.

Not to have analysed in depth the status and the role of the military in the global environmental crisis has yet another consequence: the management of environmental problems is seen by the Brundtland Commission in very similar terms as the management of military and political conflicts. Though the Commission notes that 'there are, of course, no military solutions to environmental insecurity',[36] it nevertheless proposes to deal with environmental insecurity in the very same way as international conflicts have historically been dealt with, i.e. through the 'joint management and multilateral procedures and mechanisms'.[37] This approach – sometimes also called cooperative management, among nation-states of course – is how the Commission proposes to deal with the environment as a security issue. The Commission says:

It would be highly desirable if the appropriate international organizations, including appropriate UN bodies and regional organizations, were to pool their resources – and draw on the most sophisticated surveillance technology available – to establish a reliable early warning system for environmental risk and conflict. Such a system would monitor indicators of risk and potential disputes, such as soil erosion, growth in regional migration, and uses of commons that are approaching the thresholds of sustainability. The organizations would also offer their services for helping the respective countries to establish principles and institutions for joint management.[38]

In short, not properly analysing the military leads the Brundtland Commission to propose a military kind of international management of environmental problems and resources, the so-called commons.

THE COMMONS

Potentially, for the Brundtland Commission, the commons include all the planet's resources, since these are in common to all people and do not just belong to nation-states. However, the idea of the commons is thought of from the perspective of what nation-states currently do manage in common, i.e. deep oceans, Antarctica, and space. The dangers of over-exploitation of the oceans through the fishing of coastal and deep sea areas, pollution from toxic dumping or run-off from land-based development into the oceans, and careless disposal of nuclear waste in the space orbits are expounded by the report. The importance of international cooperation is stressed and the dangers of national self-interest is cautioned against.

However, the traditional meaning of the term 'commons' is quite different from the meaning the Brundtland Commission assigns to it.[39] The commons are usually managed by people – not nation-states – at a local and not at a global level. The commons are providing livelihoods for the people directly managing them. Basically, the commons refer to traditional communities who own their resources jointly and distribute their wealth wisely. By referring to the same term, 'the commons', the Brundtland Commission wants to make us believe that the planet as a whole can be managed in the very same way. However, this use of the term should not make us forget that the Commission effects two fundamental transformations here. First, the global commons are, in its view, no longer managed by traditional communities and their members, but by nation-states. Second, managing the global commons is not about the wise use of the wealth locally produced. Rather, global management of the commons is simultaneously resource and risk management.

The idea of global management hands over the policing of the commons and their sustainable development to a global establishment, its institutions and agreements. Global management means global policing and therefore a militaristic model of fighting for 'freer' and more 'competitive' markets that will supposedly distribute things more equitably without examining the inherent nature of enclosure, export, and community destruction in these methods. Quite logically, the recommendations of the Commission for legal and institutional change all pertain to global resources and risk management. Community groups have received little support from the Commission apart from that given to NGOs. But this support is interesting, because it mentions them mostly in the context of their ability to reach groups that government agencies cannot and taking on jobs that need to be done. Note that the orientation is top-down: priority goes to governmental and international institutions and when they are not able to solve the problems from above, NGOs are given this task. The idea that community groups might know more than governments, and that they might be better suited to support the commons, is not considered.

CONCLUSION

To sum up the discussion of the Brundtland report we can conclude that the Commission basically reformulates the by-now old development myth, i.e. the myth of unlimited industrial development. It is the old idea of stages in the development process where, in a first stage, a given society draws from its natural resource base in order to build up its own intellectual, economic, and technological capacities. The second stage of development is then said to draw upon these capacities, rather than on the natural resources base. This model is based on the idea that gradually a society can make itself become independent of nature. 'Sustainable development', then, is just another word for an economic process that is drawing on a society's techno-economic capacities, rather than on the natural resources base. Of course, there are 'no limits to growth' in this model, given that the more developed a society is, the less it depends on resources that are external to it, i.e. the more it can develop sustainably. It is with this prospect of achieving independence from nature that most natural and engineering sciences are developed. And it is with the complementary prospect of optimizing a society's management capacity to sustain such development that the social sciences are pushed forward. Therefore, sustainable development becomes a matter of financial and human capital, technology, and organizational capacity. If some societies have not achieved sustainable development yet, so goes the argument, it is basically because they lack the financial, human, technological, and organizational capacity to do so. If other, more developed, societies do not do well in terms of sustainable development, this, so goes the argument, is due to a lack of economic, technological, and organizational efficiency.

So far the discourse of the Brundtland Commission is, therefore, hardly new. The only new element is that development is now looked at from a planetary or global perspective. Instead of stressing the development of a given society or country, the stress is now on the development of the planet as a whole. In that sense, the Brundtland Commission has succeeded where GATT has failed. It has managed to make the development discourse universal. One of the key tools in doing so has been the ambiguous use of the term 'commons' or 'global commons'. From the Commission's perspective, the commons are the natural resources available planet-wide. These resources are needed in order to move societies to the second stage of industrial development, i.e. to sustainable development. Also, looking at these resources on a planetary scale, the Commission at least implicitly admits a certain finiteness of these resources. However, the Commission still thinks that the major limits to growth are not the natural resources, but the state of technology and social organization. Output limits – such as pollution – are only of interest to the Commission if they risk damaging the resource base. Says the Brundtland report:

The concept of sustainable development does imply limits – not absolute limits but limitations imposed by the present state of technology and social organization on environmental resources and by the ability of the biosphere to absorb the effects of human activity. But technology and social organization can both be improved to make way for a new era of economic growth.[40]

The commons are, in the eyes of the Commission, the natural resources on which all societies need to draw in order to get them to the second stage of sustainable development. Using the idea of the commons in this global context is, as we have shown, a perversion of the original meaning of the term. The commons are the common land used by a local community for activities that benefit the entire community. Commons were therefore managed by the community. Referring to the planetary resource base in terms of 'commons' suggests that the 'human community' is to manage these commons. However, the crux is that on a global level the Commission is not thinking of the community of individuals, but of the community of nation-states. The Commission refers to the oceans, space, and Antarctica as examples of a common management of common resources, as well as risks affecting these resources. Implicitly, however, the Commission thinks that all resources should be managed in common, i.e. between nation-states. Note the shift from communities of individuals managing their commons to the community of states managing the global commons.

In short, the Brundtland report strengthens the old development discourse by lifting it to a planetary imperative. As in the original development paradigm, sustainable development – which is but another term for 'modernity' – is to be achieved in the second stage of development. From a planetary perspective, which is the only novelty the Brundtland report proposes, we are currently in transition from the first stage (pillage of natural resources or pre-modernity) to the second stage (sustainable development or modernity). This process, the Commission says, has to be managed on a global scale, and its managers are the existing nation-states.

2

SOUTHERN ELITES

The report of the South Commission entitled *The Challenge to the South* is another important document to put in the context of the UNCED process. Although the origin of the South Commission is unrelated to that process, the political positions articulated in the report became increasingly important as the UNCED negotiations unfolded. The Commission was set up in 1987 by the non-aligned movement on the initiative of Malaysian Prime Minister Mahathir Mohamad and headed by the former President of Tanzania, Julius Nyerere. Its role was threefold: to investigate the common problems of the Southern countries; to examine the possibilities of their working together to solve these problems; and to develop a new dialogue with the North.

It was set up at just about the same time that the Brundtland Commission delivered its report, but was originally concerned with very different problems. Indeed, 'for most countries of the South, the decade of the 1980s came to be regarded as a lost decade for development'.[1] The South Commission identified in its report various aspects of this 'development crisis of the 1980s'. In the beginning of the 1980s economic activity in the industrialized countries slowed down and reduced the demand for imports from the South. Also, the 'debt-related transfers, normally from North to South, were reversed and became a major drain on Southern economies as from 1984'.[2] This was, moreover, aggravated by the fact that almost all commodity prices fell in the second half of the decade. Finally, 'direct foreign investment in developing countries fell by about two thirds in real terms between 1982 and 1985'.[3]

As a result, and after the development decade of the 1970s, many developing countries experienced a crisis in the 1980s, not to mention the loss of many illusions associated with the perspective of future development. On top of that, the end of the Cold War did not lead to renewed interest from the North in the South. Rather, a situation arose where 'both attention and technical and financial resources are being directed from development in the South to economic reconstruction of Eastern Europe'.[4] In short, the South Commission emerged in the context of the betrayal of the hopes the South had started to nourish in the 1970s. The 'crisis of development' is, therefore, above all a crisis of the perspective of further development.

The environment is not really part of the considerations of the South Commission. In fact, of the 300-page report, only a few pages are devoted to environmental issues and problems. At the Least Developed Countries (LDC) meeting in Paris in September 1990 of the 42 poorest countries of the world, where the Commission presented the report, a Bangladeshi diplomat told a press conference that they could not be bothered about the environment when their people were starving to death. However, as the UNCED negotiations got more serious in 1991, Southern governments stopped complaining that the environment was a luxury of the rich, and started to get their act together. Perhaps they began either to understand that food security, as the Brundtland Commission had pointed out, was also an environmental issue, or they realized that the environment was a bargaining chip for the South, an issue around which they could rally and demand more help from the North.

As a result, the South Center (established in Geneva at the South Commission's final meeting in October 1990) published a 20-page brochure entitled 'Environment and Development: Towards a Common Strategy of the South in the UNCED Negotiations and Beyond'.[5] As we will see later, this brochure is more specific about the environment than is the South Commission's report. Yet, it is above all a strategy paper for the Southern governments in order to get the North to give further support to Southern industrial development, using Northern environmental concern as leverage.

To sum up, the South Commission's main and almost only concern is industrial development and economic growth. This is not very surprising, seeing that almost all of the Commission's 29 members have, at some point in their lives, been either economics professors or ministers of economic planning and development, or both. Needless to say, the South Commission is basically a reflection of the Southern countries' most Westernized élites. It comes as no surprise that these Southern élites, as represented in the

Commission, are most interested in the pursuit of development. They look at it from a national, from a South–South, and from a North–South perspective, a distinction we will follow here.

ECONOMIC GROWTH – THE NATIONAL PERSPECTIVE

When it comes to development – and that is all their report is about – the South Commission is at least clear: no need to use such ambiguous terms as 'sustainable development'. Instead, the Commission talks about 'sustained development' and 'economic growth', for which all national resources, including women, must be 'mobilized' – note the military language. For the South such growth is said to be 'imperative'. The basic unit for achieving it remains the nation-state. Indeed, the state's role in promoting economic growth and development is given much thought in the report. The South Commission seeks to make clear that there is a need for state intervention in order to promote the capacity-building that is normally neglected by the market, such as education and scientific research and development. It also points out that many Southern states have administrative systems that were set up by their former colonial masters in order to serve their – the colonists – best interests. And it makes some very valuable suggestions on the importance of rethinking the state.

However, all this serves the purpose of making the state more fit to be a development agent. It is from this perspective that the report encourages public participation as well as scientific research and development. People need to be 'mobilized' in order to participate actively in the national development endeavour. Appropriate political structures – such as democracy and public participation – have to be allowed in order to promote economic growth. Says the Commission:

Development can be achieved only if a nation's people – its farmers, workers, artisans, traders, businessmen, entrepreneurs, and public officials – are able to use their energies creatively and discharge their functions effectively. This in turn is critically dependent on the establishment of efficient institutional mechanisms – both private and public – that enable all economic actors to play their roles.[6]

And appropriate political reforms – such as land reforms – have to be allowed as well: 'Land reforms leading to more equitable patterns of ownership and more efficient land use are indispensable for increasing agricultural production and food security'.[7] The same idea of 'mobilizing civil society for development' applies also to women. Says the Commission: 'The mobilization of women as equal partners in all development processes therefore needs priority attention of policymakers'.[8]

In all fairness it must be said that the report does state that 'development should be consistent with the evolving culture of the people'.[9] However, if one examines in more detail what the Commission means by 'culture', one finds a very Western definition, namely one where culture has no relationship with nature, i.e. it is conceived basically as a luxury, a form of 'collective entertainment'.[10] If the Commission is interested in culture at all, this is, above all, because 'cultural values can produce social reactions, from apathy to hostility, that hinder efforts to implement development strategies'.[11] Therefore, not surprisingly, 'development strategies . . . must include as a goal the development of culture itself'.[12] Although it is not said explicitly in the report, popular cultures will have to evolve towards a 'scientific culture' if economic growth is to be achieved successfully in the South.

The Commission does, however, say explicitly that the adoption of Northern, Western, and modern science must be a stated goal of any development strategy: 'The creation, mastery, and utilization of modern science and technology are basic achievements that distinguish the advanced from the backward world, the North from the South . . . . Thus, future development policies will need to address with great vigor the closing of the knowledge gap with the North'.[13] Therefore, there is an urgent need for so-called 'human resources development' and 'capacity-building'. Says the Commission: 'Progress in this field calls for the overhaul of educational systems, in order that more attention may be given to education in science and to training in engineering and technical skills'.[14]

Capacity-building, democratization and political reforms are all seen by the Commission as necessary prerequisites in order to embark on the path of national economic growth. This is especially true in these difficult times when such economic growth in the South can no longer be expected to result automatically from the economic growth in the North through a trickle-down effect, nor from Northern aid, given the East–West rivalry. Such 'self-reliant and people-centered' – substitute 'national' – development will have to focus, according to the Commission, on four areas, namely agricultural development and food security, industrialization, service industries, and trade strategies.

The Commission does analyse the complicated and increasingly difficult trade situation for the South. Indeed, the South has always been at a disadvantage in international trade treaty discussions, for example in places like the GATT talks where it has few negotiators and often no expertise at all. Also, both countries and multinationals in the North are undergoing further consolidation into the unified North American and European Community trading blocs that will strengthen their producers while impeding Southern products, and making life very difficult for Southern producers unless they band together in a similar fashion. But despite these observations, the development ideology prevails over common sense, and the Commission concludes that international trade – together with fast and strong national economic growth – is the main 'tool of progress'.

In the same way as for the North, industrialization is a key part of the Commission's development strategy. It recommends that attention be paid to economic efficiency and technological dynamism. Once again, proper incentives, subsidies, and taxes are discussed in some detail. The Commission is also in favour of Southern countries' taking advantage of the new and fast growing service industries like tourism and finance.

Under the heading of agricultural development and food security, the South Commission discusses the issue of unequal distribution of land and the dumping of cheap food from Northern countries. But, like the Brundtland Commission, it largely ignores the fact that the best land in the South is often engaged in producing export crops, although it is critical of government policies that do not promote food production for local consumption and those that encourage the consumption of imported foods. For Africa, the report says, new crops need to be found that will suit the fragile soil. Particular attention needs to be paid to post-harvest storage methods to avoid the 40 per cent loss at that stage. Overall, the message of the Commission in this matter can be summarized as the 'industrialization of agriculture'. Says the Commission: 'Particular importance needs to be attached to industry's link with agriculture. The rapid expansion of the cultivation of food crops can be facilitated by industrialization'.[15] This is not surprising, since in the Commission's eyes the environmentally and culturally destructive Green Revolution has actually been a success to be replicated: 'The successful achievement by the Green revolution in Asia has lessons for countries with sluggish agricultural growth'.[16]

It is only in connection with security and agricultural development that the environment is actually mentioned by the Commission. Just as for the Brundtland Commission, the environment is basically an economic resource. As such it has to be rationally managed, while being further exploited. Says the Commission: 'The countries of the South will need to make a concerted effort to counteract environmental stress, as sustained development will require preservation and development of natural resources, as well as their rational exploitation'.[17] Overall, the message is that 'the South has no alternative but to pursue a path of rapid economic growth, and hence to industrialize'.[18] And the Commission insists: 'This [industrialization and pollution] is just, as well as necessary, given the enormous disparity in the levels of energy consumption between the North and the South, and the indisputable right of the South to develop rapidly to improve the well-being of its people'.[19]

SOUTH—SOUTH COOPERATION

This part of the South Commission's report is perhaps a more significant contribution to the political debate than the previous part on sustaining national industrial development. Indeed, all the Commission says about development hardly breaks new ground. Almost all of it has been examined at length by economists and other development thinkers over the past twenty years. Much effort has been made over the years to implement these new policies, generally with disastrous consequences for the people, their cultures, and the environment, local and global.

Southern cooperation too is not a new idea, but has hardly ever been implemented in any serious manner. So the South Commission's call for strong collective action, such as the need for the South to speak together on issues of common concern like debt, is still welcome. However, it is disappointing to see that much of this discussion centres on creating Northern style institutions in the South. Yet this is not really surprising, as Northern style institutions are probably best suited to further the Northern style development the Commission seeks to promote.

The Commission starts off by explaining the rigidities of a world organized along North–South lines where all the trading routes are directed northwards. It notes the phenomenal economic success of oil production agreements and argues that this should be extended to more of the commodities that the South exports. Unfortunately, it hardly mentions the fact that UNCTAD has been working on this issue for decades with a singular lack of success, mainly because of the subordinate role that it plays to more powerful institutions like GATT which have exactly the opposite interests. Attempts to shore up the prices of several commodities like coffee and rubber by controlling production have met with little success. UN institutions have also not had much success helping countries diversify production after the collapse of prices of major exports. Nor does the Commission address contradictions to its supposed concern for the environment that may arise through creating a major demand for importing toxic waste to the South, for example.

The Commission goes on to explain the necessity of creating Southern institutions that will ensure Southern cooperation in a variety of areas. For example, it thinks that it is very important to create Southern multinationals and a South Bank that will take on the role of the World Bank, but for the South. Multinationals and the World Bank are possibly two of the worst examples of Northern development strategy, as they are two of the biggest contributors to cultural and environmental destruction in the South in recent years. It would be particularly disastrous to ape these institutions as part of a strategy for South–South cooperation. More valuable ideas include Southern institutions that would gather and exchange Southern knowledge such as a proposed South Secretariat to organize Southern countries to speak with a collective voice, the proposal for regional groups to help settle regional conflicts, and the recombination to strengthen existing institutions like the Third World Academy of Sciences. Perhaps the South Center, created in 1990 as the follow-up to the South Commission, is meant to be a step in this direction. But why, then, is the South Center located in Geneva, Switzerland? And then again, the South Commission does not have any particular concern for traditional knowledge systems and local communities. All the suggestions for the creation or support of Southern institutions are in fact directed at copying Northern science, technology, education, and institutions in order to boost trade and economic growth in the South.

NORTH—SOUTH RELATIONS

Quite logically, since the South and the North are basically aspiring to and competing for the same goals, their relationship is portrayed by the South Commission in terms of conflict, especially, as we shall see, when it comes to the environment. Of course, the Commission does call for debt cancellation in the vain hope that Northern multilateral and private banks will heed its words. Like the Brundtland Commission, the South Commission also calls for more loans as a way to build up the infrastructure of Southern countries, emphasizing, as always, the need for the North to donate a minimum amount of their national income. Once again, it is a statement that is, in our opinion, hard to justify when it has become obvious over the years that more loans will lead to more debt and, as a result, more environmental destruction. The Commission also calls for more multinational investment as a way for Southern countries to receive new business and technological skills. This is also a position that is hard to justify when the Commission has spent such a lot of time explaining that increased dependency on the North has led more and more to the South being exploited by the North. To be fair, the Commission does attempt to give this some balance by saying that foreign investment by multinationals needs to be monitored for its impact on the South. But it is not clear from the report who is going to do this monitoring.

Finally, the Commission also says that disarmament is an area that could open up financing for the South, suggesting that part of the money saved by a reduction in military budgets could be used to help meet Southern technological needs. But the issue of militarization and disarmament stands out largely by its omission. At the very beginning of the report the Commission notes that Southern countries spend a large portion of their budgets on the military. At no point does it mention the major role that Northern aid plays in this, and at no point does it stress the need for disarmament in the South. Instead, it calls mostly for new security arrangements in the South and stresses the need for regional solutions to regional conflicts. Had the Commission condemned the exploitation of people and the destruction of the environment by both the North and the South through military spending, and called for the regulation of the industries that supply military hardware, it could have made a much stronger case for an alternative to current development strategies.

The major novelty in terms of North—South relations is probably to be found in the environmental field, triggered by global environmental concerns and the UNCED process. In a very interesting briefing paper – written in 1991 by the South Center for the Southern governments at the UNCED negotiations – the South Commission's thinking is translated into an environment and development negotiation strategy. In this paper, the global environment is now perceived as a limited pie of pollution rights, to which both the South and the North aspire. Given, as outlined above, the South's quest for and perceived right to industrial development, the negotiation strategy of the Southern governments therefore must be to ensure that the South 'has adequate "environmental space" for its future development',[20] environmental space meaning the right to destroy and pollute the global environment further. Therefore, one of the key issues for the South in the UNCED negotiations 'is to indicate clearly the areas where it expects the North to adjust its production and consumption patterns in such a way as to leave the South with adequate environmental space for its development'.[21]

But although the South Center, as the South Commission before it, clearly sees the North and South competing for environmental space for their respective industrial development, thus articulating the new North–South conflict, we must not forget that this conflict stems from the fact that both share exactly the same aspirations and ideology of industrial development. As a matter of fact, the South Commission is much clearer in articulating this ideology than is the Brundtland Commission: development is an imperative for the South, it says, 'since only rapid industrial development can create the resources to satisfy the basic requirements of their populations'.[22] Moreover, the Southern nations must mobilize their people for that purpose and organize themselves in order to get the maximum out of the North. With the environment and in the context of the UNCED process, the Southern élites once again seem to have found a substantial bargaining point to get the North to support their industrial development. But clearly the environment is none of their concern. Like for the Brundtland Commission, it is a resource and 'space' for further industrial development.

3

RIO AND BUST

At Rio de Janeiro in June 1992 most heads of government signed a package of agreements, namely a biodiversity convention, a climate change convention, a statement on forest principles, an agreement to work towards a desertification convention, the Rio Declaration on environment and development, and Agenda 21, a mammoth 800-page plan for saving the planet in the twenty-first century. In this chapter we briefly present and critically discuss these documents.

However, to begin with, it should be made clear that the main message and content of all these documents is above all a consolidation of the type of thinking we have identified in the reports of the Brundtland and the South Commissions. In other words, none of the documents displays any new or original way of looking at environmental and developmental issues. Before looking at each of the documents in more detail, it will be useful to recall what all of them missed.

WHAT WAS MISSING?

What all of them missed was summarized, in our view, in a '10-point plan to save the Earth Summit' sponsored by Greenpeace International, the Forum of Brazilian NGOs, Friends of the Earth International, and the Third World Network (a coalition of Southern NGOs).[1] This plan was presented in Rio and endorsed by over fifty other NGOs. The plan called on the Earth Summit to achieve the following:

1. Legally binding targets and timetables for reduction in greenhouse gas emissions, with industrialized countries leading the way.
2. A cut in Northern resource consumption and transformation of technology to create ecological sustainability.
3. Global economic reform to reverse the South–North flow of resources, improve the South's terms of trade and reduce its debt burden.
4. An end to the World Bank control of the Global Environmental Facility (GEF).
5. Strong international regulation of transnational corporations, plus the restoration of the UN Center on Transnational Corporations, rather than allowing the Business Council for Sustainable Development to go unopposed in the UNCED process.
6. A ban on exports of hazardous wastes and on dirty industries.
7. Address the real causes of the forest destruction, since planting trees, as UNCED proposes, cannot be a substitute for saving existing natural forests and the cultures that live in them.
8. An end to nuclear weapons testing, phase-out of nuclear power plants and a transition to renewable energy.
9. Binding safety measures – including a code of conduct – for biotechnology.
10. Reconciliation of trade with environmental protection, ensuring that free trade is not endorsed as the key to achieving sustainable development.

Neither Northern consumption, nor global economic reform, nor the role of transnational corporations, nor nuclear energy, nor the dangers of biotechnology were addressed in Rio, not to mention the fact that the military was totally left off the agenda. Instead, free trade and its promoters came to be seen as the solution to the global ecological crisis. This is, in part, due to the fact that the underlying documents, in particular the Brundtland report, on which the thinking in these documents is based, were flawed to begin with. As we now examine in more detail each of the Rio documents, we refer to these omissions and flaws. Some of them are addressed in this chapter, whereas others are discussed later in the book. Points 7 and 9 relate to the biodiversity convention that we discuss next, while point 1 relates to the climate change convention that follows. Transnational corporations are discussed in detail in

Part III, as is the issue of resource consumption mentioned in point 2. Point 4, about the World Bank, is discussed in Part IV. We have already discussed some issues raised in point 2 (consumption and resources) and point 3 (terms of trade and debt), as well as those in point 10 (free trade). The two other points, on international waste trade (point 6) and nuclear power (point 8), are discussed briefly at the end of this chapter.

THE BIODIVERSITY CONVENTION

The biodiversity convention, like the climate change convention, was actually negotiated separately from the UNCED process in a so-called international negotiating committee (INC). Though both INCs met at different times to the UNCED PrepComs and had different national representatives, the issues, the stakes, and the conflicts turned out to be very similar to those that congregated around the same questions in Agenda 21. This was particularly the case after these two negotiations were declared part of the UNCED process.

As a matter of fact, when the negotiations for such a convention were initiated by the United Nations Environment Programme (UNEP) in 1990, the biodiversity convention was to become a separate convention, and nobody planned that it would be ready to be signed in June 1992. The origin of the convention goes back to the concern for the destruction of the tropical rainforest, voiced mainly by Northern conservationist NGOs, in particular IUCN and WWF. These NGOs, in collaboration with the World Resources Institute (WRI), the World Bank, and UNEP, drafted the original texts. They mainly dealt with the protection and conservation of biodiversity along a quite traditional, i.e. resource management, approach. It was only during PrepCom III in August 1991 in Geneva that the so-called 'Group of 77' (G-77) which now comprised most developing countries (i.e. currently 128 countries) asked that the issue of biotechnology be included in the convention on biodiversity. Finally, a compromise document was drawn up in Nairobi a month before Rio and taken to the Summit. Negotiations had, in fact, been delayed because the USA had demanded substantial changes, which they got. Despite these concessions, the USA refused to sign the convention in Rio.

According to Patrick McCully:

It is likely that the biodiversity convention with its legal intricacies and obscure language would have received little attention at Rio were it not for George Bush's — US president at that time — refusal to sign it. The US's intransigence on this issue became a focus for NGOs' demonstrations and press interest, the anger against Bush implying that the convention was somehow going to mark a great advance.[2]

Indeed, despite many critical notes from some NGOs, the biodiversity convention was generally considered the biggest success of the entire UNCED process. A total of 156 countries signed the convention in Rio, and four more have signed it since, but only six had ratified it as of February 1993. Many NGOs were indeed keen for the convention to be signed. While recognizing that it had many shortcomings, groups at a workshop held by the Brazilian NGO SOS Mata Atlantica (SOS Atlantic Forest) called the convention 'a milestone in an ongoing process for the conservation and wise use of the world's biodiversity'. The Third World Network, however, noting that last minute changes had been made to the convention's provisions on ownership of genetic resources, advised Southern countries not to sign.[3] Amid the excitement of the Summit and the general anger with the USA's pro-business stance, few noticed the warning.

Indeed, the biodiversity convention is just one of many typical examples where the concern for exponential destruction of the world's biodiversity has been perverted into a preoccupation with new scientific and (bio-)technological developments to boost economic growth. Or as Vandana Shiva puts it: 'It is ironical that a convention for the protection of biodiversity has been distorted into a convention to exploit it'.[4] Though this is not surprising, given the conceptual framework of the Brundtland Commission discussed earlier, it is nevertheless worth while identifying the three key arguments that cement this perversion: first, the convention gives 'nation-states the sovereign right to exploit their own resources pursuant to their environmental policies',[5] thus transforming biological diversity into a natural resource to be exploited and manipulated. Then, the convention implicitly equates the diversity of life — animals and plants — to the diversity of genetic codes, for which read genetic resources. By doing so, diversity becomes something modern science can manipulate. Finally, the convention promotes biotechnology as being 'essential for the conservation and sustainable use of biodiversity'.[6]

Not surprisingly, biotechnology, according to Agenda 21, is not only good for the conservation of biodiversity, but it is also good for improving agricultural production as it will 'increase the yield of major crops, livestock, and acquaculture species'.[7] The biotechnology industry is therefore beneficial to humanity in at least two ways, i.e. to conserve biodiversity on the one hand and to improve production on the other. Interestingly, as Third World Resurgence has pointed out,[8] the thrust of the biodiversity convention – as well as that of chapter 16 of Agenda 21 dealing with biodiversity and biotechnology – is exactly the same as a document prepared by the biotechnology industry for UNCED. In it the International Biodiversity Forum says that genetically manipulated organisms are 'natural' while at the same time claiming that they are improvements upon nature due to 'increased efficiency'. The document, like the convention, also says that (1) 'modern biotechnology will help maintain biodiversity and ensure genetic diversity', and that (2) 'biotechnology will . . . [provide] extensive environmental benefits for sustainable growth'.

To recap, the main stake raised by the biodiversity convention is the issue of ownership and control over biological diversity. In the case of the North, and the USA in particular, the major concern was protecting the pharmaceutical and emerging biotechnology industries, which get their raw material from forests. In the case of the South, the concern was mostly ensuring that governments and industries could continue to exploit their own natural resources. Obviously, the convention is a compromise with considerable advantages for the North. Who spoke and speaks for the local communities who often sustain and depend on biodiversity for foods, medicine, and their way of life? Once the biodiversity convention had included the question of biotechnology – as demanded by the South – and which subsequently become a propaganda instrument for the biotechnology industry, the main debate between the North and the South was over patent rights, redistribution of profits from biotechnological production, access rights and control over genetic banks, as well as debates about the safety of biotechnology. Yet, amid these financial and political controversies the main issue was forgotten, namely the identification of the main causes of the destruction of biodiversity and the drawing up of action plans to address these causes.

THE FRAMEWORK CONVENTION ON CLIMATE CHANGE

The negotiations for the climate convention are a good example of what happens if a global environmental problem cannot be turned – unlike the case of biodiversity – into the promotion of further industrial development. Therefore, the climate negotiations are probably best characterized as an 'effort to avoid conflicting positions through vagueness and ambiguity'.[9] Like the biodiversity convention, the climate convention was negotiated separately, a process initiated by the warnings of the Intergovernmental Panel on Climate Change (IPCC), which stated in 1990 that unless emissions of greenhouse gases such as carbon dioxide were cut significantly, the world could face unprecedented global warming. Global warming would lead to rises in sea level and coastal flooding, unpredictable weather patterns, and drought, and therefore decreased agricultural productivity, and further hunger and migration. Because carbon dioxide is mostly responsible for global warming, the IPCC concluded that carbon emissions needed to be cut by 60 per cent at least in order simply to stabilize current carbon levels in the atmosphere. The IPCC had also assessed that the industrialized North accounted for the majority of carbon dioxide emissions, basically due to the fact that such emissions are totally correlated with fossil fuel consumption, fossil fuels being the primary motor of industrial development. The USA alone, the IPCC stated, accounts for about 23 per cent of worldwide carbon dioxide emissions.[10]

After the negotiations on emissions reduction started in the International Negotiating Committee on Climate Change in 1990, it rapidly became clear that at best governments would put pressure on their industries and other greenhouse gas emitters to return to 1990 levels by the year 2000, a figure that would not come anywhere near the 60 per cent reduction that the IPCC scientists had called for. But the USA rapidly caused deadlock by refusing to set a target for even stabilizing, let alone reducing, emissions of carbon dioxide, because it said it would cause a major setback for its economy. All the other OECD countries had agreed to go along with the 1990 target. But at the final INC meeting in April 1992 in New York, everybody bowed to US pressure. Previously, a US government Commission had concluded that the USA could actually adapt to and mitigate the consequences of climate change, and furthermore win a strategic advantage by doing this.[11] As a result, the convention contains no legally binding commitments for industrialized countries to stabilize, let alone to reduce, carbon dioxide emissions.

The Japanese and the European governments who had condemned the USA for its stand, some of whom even went to Rio with a counter-proposal to steal the thunder from the USA on the final day, quietly shelved their plans too. And President Bush bought himself some publicity by offering to contribute US$150 million over the next two years for Southern countries to figure out how they could cut their greenhouse gas emissions. This proposal, of course, is ridiculous given the fact that the USA is estimated to be the source of a quarter of the world's greenhouse gas emissions and therefore the best starting point for reduction.

The convention — which is indeed a framework and not a real convention — now only requires that Northern countries submit a list of their plans for carbon dioxide reductions to the secretariat of the convention and report on its follow-up. They are also 'encouraged' to take up joint ventures with other countries; by planting trees in other countries which help absorb greenhouse gases, for example, they can take credit while not cutting back on their own emissions. Northern countries are also supposed to come up with money to help cut emissions in the South and assist in the transfer of relevant technology to help Southern countries in these efforts. Once again, no actual amounts of money or precise commitments of any kind were made. Finally, Southern countries are asked to submit an inventory of their greenhouse gas emissions and plans to reduce them.

This toothless framework convention had been signed, as of February 1993, by 155 countries and ratified by eleven of them. It will come into force as soon as fifty of them have ratified it. A secretariat to monitor the follow-up of the convention has been set up in Geneva, just as the secretariat to follow up on the biodiversity convention is also located in Geneva, the 'environmental capital of the world', as the Swiss like to think of it. The IPCC continues to operate under the joint sponsorship of UNEP and the World Meteorological Organization (WMO).

Yet the issue of climate change is actually quite simple compared to other global environmental problems, as primarily the fossil-fuel based industrial development needs to be slowed down. However, this is acceptable neither to governments nor to any other organization whose primary mission is to promote economic growth. None was willing to go beyond the amount of carbon dioxide reductions that technological progress made possible thanks to efficiency gains. As a matter of fact, the OECD countries which were advocating such a reduction hoped to gain a competitive economic advantage from the new technological developments this would have spurred. But no one, either in the North or in the South, was and still is willing to cut into industrial development. Therefore, even the stated objective of the convention maintains that economic development is the ultimate goal and that ecosystems will have to adapt. The objective simply is to stabilize carbon dioxide emissions so that these ecosystems will have more time to adapt. The convention states that greenhouse gas concentrations should be stabilized 'within a time frame sufficient to allow ecosystems to adapt naturally . . . to ensure that food production is not threatened and to enable economic development to proceed in a sustainable manner'.[12]

THE AGREEMENT ON FOREST PRINCIPLES

In 1990 a group of six countries in the North – Canada, France, Germany, Italy, the United Kingdom, and the United States – asked that a third convention be negotiated on protecting the world's forests. They had in mind, in particular, the world's tropical rainforests, which are currently vanishing at an estimated 17 million hectares a year, and are considered by the North as valuable sinks for greenhouse gases. What is often forgotten in this equation is that the Northern forests were also sinks for greenhouse gases before they were cut down. Moreover, many Northern countries are also planning to cut down their forests. For example, Canada and Russia are currently cutting down the world's remaining boreal forests. Siberia alone contains more forests than the Brazilian Amazon and since Russia turned capitalist it has been signing logging agreements with corporations in virtually every Northern country.

The talks on the forest convention broke down after the South refused to give in to what it called a possible infringement on national sovereignty. Malaysia was the major campaigner on behalf of Southern governments. They argued that the convention on the protection of forests would jeopardize their rights to their own resources. This fight in fact was not resolved and a forest convention was postponed to the indefinite future. Instead, the governments in Rio agreed to a 'non-legally binding authoritative statement of principles for a global consensus on the management, conservation and sustainable development of all types of forests.'[13]

The forests have clearly become a new symbol of the North–South conflict, and to a certain extent they have even become a 'hostage' of that conflict.[14] The North sees tropical forests as common property, whereas the South expects financial compensation for forgoing the exploitation of its forests. As a result, the agreement goes as far as to establish every country's 'sovereign right to conversion of forests to other uses',[15] which in plain language means the right to cut forests down as one pleases. After these forest principles, forests are now clearly declared to be 'national resources', since Third World governments in particular, especially Malaysia and Indonesia, opposed any agreement which would have limited their ability to cut their own forests as quickly as they wanted and stressed their sovereign right to develop.

In any case, the forest principles, like chapter 21 of Agenda 21 which deals with forests, got it all wrong. There is no mention of a relationship between forests and diversity, which is not really surprising since, as we have seen, biodiversity has been perverted into a (bio-)technological problem. Also, despite the fact that today's No. 1 problem for the forests is deforestation, deforestation is never mentioned in these principles. UNCED does not see deforestation as something to be combated. This, again, is quite logical since in its view 'the pressure on forests results from human [sic!] activity'.[16] This implies that all humans are equally responsible for industrial logging in Brazil, Malaysia, Canada, Siberia, etc. Quite logically, then, the forest principles propose 'planting trees', and 'sustainable forest management' as an answer. But even here 'UNCED manages to ignore much of the forestry development literature of at least a decade or so, especially in the areas of social and community forestry'.[17] In short, the statement on forest principles is more than a step backwards. It is 'a step towards further legitimizing the policies of those actors – transnationals, multilateral development banks, UN agencies, etc. – that have to date contributed to a large extent to the crisis of the tropical, temperate, and boreal forests'.[18]

STEPS TOWARDS A CONVENTION ON DESERTIFICATION

Upset about the Earth 'Summit's preoccupation with the Northern view of environmental conservation, African delegates in particular lobbied hard for a desertification convention to address some of their most pressing problems. Initially, however, they did not have much luck. UNEP statistics say that 35 per cent of the Earth's land surface is threatened by desertification and it has been trying for years to develop a systematic programme that would tackle this issue. In 1977 UNEP launched a plan of action, which would have been the most ambitious effort to combat desertification. But this plan fell through for lack of funds. The few programmes that were funded failed miserably, like one glaring example to reforest Northern Nigeria which failed because of lack of water, an issue the planners had apparently forgotten to look at.

Chapter 12 of Agenda 21 deals with deserts and droughts, and basically suggests better information and monitoring of desertification, soil conservation measures, and support for local programmes. The approach proposed in Agenda 21 is very heterogeneous: the Swedish foreign ministry, for example, instructed its delegation to suggest that military satellites could be used to monitor desertification. Some Southern countries opposed this because they were worried about the possible military use and control of such information. UNEP envisaged a convention that would take a more local approach to the problem than previous plans by focusing on education and public participation in desertification control, rather than paying for large 'greening' projects for the desert. Other ideas include helping farmers to abandon farming and diversify into other industries because of the pressure farming puts on the land. Overall, chapter 12 of Agenda 21 on desertification remains very abstract and is of little concrete use.

One can say that there was very little interest from non-African countries in desertification, and it was even disputed whether desertification is a global problem. But at the last moment, governments in Rio agreed to set up an inter-governmental group at the General Assembly in New York at the end of 1992, to discuss steps towards a convention on desertification. Subsequently, the UN General Assembly decided to establish an international negotiating committee on desertification – again to be located in Geneva – thus setting the negotiating process in motion. The convention should be ready to be signed in late 1994.

THE 'RIO DECLARATION'

Originally conceived as the environment and development equivalent of the Universal Declaration of Human Rights, Maurice Strong first wanted to call this document the 'Earth Charter'. At PrepCom III, the Group of 77 (G-77) decided that it did not like this name because it smacked too much of the environment and not enough of its primary concern, i.e. development. Despite the fact that the Earth Charter was probably what Maurice Strong was most attached to, G-77 prevailed and the document became watered down from a charter to a declaration on environment and development. Strong, though, has not given up. Recently he declared: 'the document must continue to evolve towards what many of us hope will be an Earth Charter that could be finally sanctioned on the 50th anniversary of the UN in 1995'.[19]

The document attempts to lay out the duties and the rights of states and peoples towards the planet. It has twenty-seven principles – there were originally supposed to be thirty-three – and officially complements the Stockholm Declaration on the Human Environment, of 1972. Like the other elements of the Rio package, it is very much the consensus product of many hours of bitter negotiations, mostly between government representatives. But its twenty-seven principles probably reflect more clearly and more concisely than any other Rio document the core philosophical assumptions and message of the entire UNCED process, i.e. basically it is a blend between the philosophy of the Brundtland report and the philosophy of the South Commission's report. As such, the Rio declaration is a document that once more reaffirms the quasi-religious belief in industrial development, seeks to mobilize all human potential and natural resources to that effect, and reasserts nation-states as the primary units to promote such development. Occasionally, it expresses concern that environmental degradation might hurt further prospects of development. But it is precisely the inclusion of such concerns that is used to justify adding the adjective 'sustainable' to the term 'development'. Let us briefly comment here on the twenty-seven principles in more detail.

The very first principle states that human beings are the centre of sustainable development concerns, a belief that is called anthropocentrism. Moreover, the entire UN system is rooted in the Western and Northern belief that only development can make human beings become more and truly 'human'. We think that this is a dangerous and short-term view. While we are not suggesting

49

that the life of a non-human is more important than that of a human being, we think that the debate 'human vs non-human' has been made irrelevant because of the new global ecological challenge. Indeed, global ecology forces us to admit that the current process of industrial development is destroying the very ecological foundations of all humans and non-humans simultaneously. Many indigenous peoples have quickly died out after the loss of their natural habitats on which they depended for food and medicines. We are currently repeating this same process at an accelerated pace and on a global scale. The death of many more other people may not be as quick but just as sure. Pretending that further development will preserve the human species, let alone make it more human, is contrary to all scientific and other indicators. We are dealing here with an institutionalized mythological belief in development which, if not abandoned, will prove fatal for humans and other species.

The second principle of the Rio declaration gives nation-states the 'sovereign right' to 'exploit' their natural resources according to their own environmental and developmental policies. It seems to us that the urgency of the global ecological crisis should have led precisely to the insight that nation-state sovereignty is obsolete and globally destructive. Instead, the entire UNCED process and the Rio Declaration in particular stress the nation-states' sovereign right to do with their environment and their people whatever they please as long as they do not harm other states.

The 'right to development' is enshrined in the third principle – with the caveat that the developmental and environmental needs of further generations be taken into account. Development is accorded a priority over the environment in the fourth principle, which asserts that environmental protection should constitute an integral part of the development process. Now, this may be a vast improvement on the ideas of just a few years ago, when environment was considered unimportant, but the thinking behind it is still conceptually flawed and wrong: the development process, it seems to us, occurs within the environment and its limits, and not the other way round. One would have expected from a declaration that supposedly marks the beginning of a new relationship between environment and development that it would get things at least conceptually right.

The fifth principle calls on states and peoples to eradicate poverty and reduce disparities in income. We are uncomfortable with this principle because it implies, as do the Brundtland Commission and the South Commission reports as well as all other UNCED documents, that poverty rather than affluence is the problem. The fifth principle is again entirely consistent with the prevailing development ideology, which wants us to believe that on spaceship Earth first class passengers and first class technology are best for everyone, including the biosphere.

We must acknowledge that the following three principles – 6, 7, and 8 – go some way to restore the balance, albeit a weak one. Principle 6 declares that developing countries, especially those that are environmentally vulnerable, should be given special priority. Principle 7 says that states have common but differentiated responsibilities to conserve, protect, and restore the Earth, pointing out that developed countries go some way towards acknowledging the pressure their societies have placed on the planet. Principle 8 calls for the reduction of unsustainable patterns of production and consumption. But we feel there is no acknowledgment from Southern élites nor developed countries regarding their pressures on the planet. There is, furthermore, no recognition of their differentiated responsibilities. We are also critical in principle 7 of the mention of the technological and financial resources of the developed countries, because it implies that these resources could substitute for other commitments to restore the Earth.

As we discuss in considerable detail in Parts III and IV, Northern financial and technological resources have been, up to now, among the major causes of environmental damage. So it comes as no surprise that principle 9 emphasizes technology transfer from North to South, especially for new and innovative technologies. Without doubt, technology transfer will be good for development. But to claim that it will make development more sustainable is, so far, an ideological and unproven assertion.

Principle 10 recommends that environmental issues are handled best with the participation of citizens at the relevant level and that they should have access to information and judicial redress – an excellent principle. Unfortunately, we also know that in the UN, the Brundtland Commission, and the South Commission the jargon word 'participation' equals the citizens' mobilization for development. We think that people should be allowed to do more than just participate in the development process. They should be allowed to make their own decisions and have their local and regional autonomy. If they chose not to be mobilized for the promotion of development, they should be allowed to do so.

Principle 11 calls for the creation of environmental legislation and standards. But it immediately gives countries a convenient backdoor by saying that standards may be 'inappropriate' – read 'too high' – for developing countries because of the social and economic costs. Not surprisingly, the counter-suggestion that development policies may be 'inappropriate' because they damage the global environment cannot be found in the Rio declaration.

Principle 12 reiterates the need for an open international economic system – read 'free trade' – and says that unilateral action – read 'trade bans' – based on environmental considerations should be avoided. This principle explicitly places UNCED within or beneath the larger context of the GATT negotiations. As above, where the environment is said to be a sub-category of development, this principle implies that environmental protection is a sub-category of global trade. In no way, it is said, should environmental protection slow down global trade. This principle is a deliberate blow against environmentalists and all other people who had campaigned for the Rio process, hoping that UNCED was the major forum to deal with the world crisis – but only to realize that the real forum where this planet's future was being decided was GATT.

Principles 13 to 15 can be called 'good but weak'. Principle 13 calls for national laws for the compensation of victims of environmental damage. Likewise, 14 calls on states to prevent the relocation and transfer of activities or substances that cause environmental damage, a good, but once again, weak principle. Principle 15 fits in well with this category by calling for the 'wide' application of the precautionary principle, which means 'don't do something if you think it may cause environmental damage' rather than waiting to see what the effects are.

Principle 16 calls on national authorities to apply the polluter pays principle, but immediately weakens this statement by saying that this should not slow down international trade and investment. Principle 17 has a similar weak element because it calls for environmental impact assessments of activities if 'significant adverse impact' is expected. Principles 18 and 19 call on states to inform other states in the event of natural disasters and activities that could have adverse environmental impacts. There is unfortunately no mention of the need to inform communities and people within a country, nor the need to ask them for their opinions on the degradation that may affect them as a result of foreign activity.

Principles 20 to 22 state the importance of women, youth, and indigenous people in sustainable development. But like principle 10, the implicit idea here is the one of mobilization for development: women, youth, and indigenous people are outside agents who have a 'vital role to play' (women), 'whose creativity, ideals, and courage should be mobilized' (youth), or who have 'knowledge' to contribute (indigenous people). None of them is treated as full agents to whom authority and decision-making power should be handed over, rather than merely being consulted.

Principle 23 says that the environmental and natural resources of people under oppression, domination or occupation should be protected. This principle was agreed on only at the last moment because Israel had made strenuous objections to it. Realizing that this could have severe implications for the occupied Palestinian lands, it agreed only on the condition that all other references to the subject be taken out of Agenda 21's chapter on freshwater resources.

Principles 24 to 26 relate to war but manage to make no mention of the military! These three principles say that warfare is inherently destructive of sustainable development. Therefore, states should make provisions for the protection of the environment in case of war (principle 24). Principle 25 states that peace, development and environmental protection are interdependent; states should resolve disputes peacefully in accordance with the Charter of the UN (principle 26). None of the three principles condemns war, not to mention the everyday destructive impacts of non-war activities on the environment caused by the mere existence of the military.

Nor are the other major environmentally destructive agents – i.e. nation-states and multinational companies – ever mentioned in the Rio declaration. Instead, it concludes with the rather lame principle 27, saying that states and people should cooperate in the development of international law to promote sustainable development. But as we see in Part II, this supposed cooperation between states and people is not as democratic as principle 27 would like us to believe.

AGENDA 21

In Nairobi, during the first PrepCom meeting in August 1991, Maurice Strong proposed a master plan, to be called Agenda 21, to put the planet on a sustainable footing in time for the next century – hence the number 21. Though non-binding for the signatory states, the idea was that, after adopting Agenda 21 in Rio, the governments would implement this master plan, or at least be inspired by it when taking environment and development related decisions.

After reviewing the thirty or so subject areas that the secretariat prepared for this first meeting in Nairobi, the government representatives then spent the bulk of the next three meetings writing the plan. When they reached Rio, the plan had forty chapters which covered the following substantive subjects: a preamble, sustainable development, poverty, consumption patterns, demographics, human health, human settlements, decision-making, atmosphere, land resources, deforestation, desertification and drought, mountains, agriculture and rural development, biodiversity, biotechnology, oceans, freshwater resources, toxic chemicals management and their transport, traffic in hazardous waste, solid waste, radioactive waste, women, children and youth, indigenous peoples, NGOs, local authorities, trade unions, business and industry, scientific communities, farmers, financial resources, technology transfer, science, education, international institutions, and legal instruments and information.

Often these negotiations seemed more like a giant editing session with 150 editors squabbling over each and every word, or perhaps more accurately 150 lawyers haggling over a settlement. The end result was an 800-page document that is quite indigestible and impossible to implement. Therefore, since Rio several people have tried to produce at least a plain language version of Agenda 21.[20]

It is impossible for us to comment on each of the forty chapters of Agenda 21. Because of the redundancy and multiple repetitions in the text this is actually not desirable either. Moreover, the overall thrust of Agenda 21 is identical to the thrust of the Rio declaration which in turn is a blend of the ideologies of the Brundtland and South Commission reports. For its critique we can refer to the above section on the Rio declaration. In this section, we would like to look critically at the six main themes of Agenda 21, identify the main aspects missing in it, and briefly talk about its possible implementation.

Daniel Sitarz has, to our mind accurately, identified the six main themes that Agenda 21 contains, namely the theme of the quality of life on Earth, the efficient use of the Earth's natural resources, the protection of our global commons, the management of human settlements, chemicals and the management of waste, and sustainable economic growth. We will follow here his distinction.[21]

THE QUALITY OF LIFE ON EARTH

Agenda 21 starts with a series of chapters dealing with the promotion of the quality of life on Earth. Most of them are composed of lofty statements the UN has been professing since its inception, such as the eradication of poverty worldwide, raising the level of general health, full employment, controlling population growth, etc. Though we have no problem with these lofty statements, in the eyes of the UN their achievement is only possible through further economic growth, not realizing that economic growth has caused the present global environment and development crisis to begin with. Of course, Agenda 21, like all other UNCED documents, never explicitly identifies economic growth and industrial development as being a problem for the biosphere and therefore for humanity. At best, Agenda 21 criticizes 'human activities' or 'current resource consumption patterns' for being responsible for the present crisis. Note that by doing so, the responsibility for the current crisis is being diluted and shifted from the major polluters and promoters of industrial development and economic growth to all inhabitants of the planet. Indeed, it is the individuals who are blamed for the current crisis: they should bear, it is argued, the main responsibility, change their own human activities, and alter their consumption patterns. Of course, we are not opposed to profound changes in individual behaviour, but we believe that such profound individual changes are only possible when paralleled by equally profound changes in the system.

THE EFFICIENT USE OF THE EARTH'S NATURAL RESOURCES

The second major theme in Agenda 21 is the 'efficient use of the Earth's natural resources'. Though, again, nobody would oppose a more efficient use of the Earth's natural resources, the crucial point here is that Agenda 21 sees efficiency as the single most important solution and thus transforms everything on this planet into a resource and attaches economic value to it.[22] Says Sitarz in his comment on Agenda 21: 'The carrying capacity of the Earth must be valued as an economic resource, if it is to be assured of protection.'[23] In the name of environmental protection, therefore, Agenda 21 extends the economic rationality to the most remote corners of the Earth, and to every single as yet untouched plant, animal, indigenous person, or gene, and feeds them back into this overall masterplan promoting (sustainable) development. Agriculture and farmers, for example, must be mobilized, says Agenda 21, for global food production. If necessary, genetically modified species must be introduced into the agricultural systems. 'Mobilization' – a military term – is actually highly appropriate, since Agenda 21 considers food production to be a 'security issue' (food security).

SUSTAINABLE ECONOMIC GROWTH

The next major theme of Agenda 21 – sustainable economic growth – is, as we have seen, the major thrust of the entire UNCED exercise. Economic growth is the objective, and the challenge is to integrate environmental protection into this objective, not the other way round. Agenda 21 sees this integration as an 'economic transition'. If successful, 'the protection of the environment will be given a proper place in the market economy of the world'.[24]

THE PROTECTION OF OUR GLOBAL COMMONS

The fourth major theme – the protection of our global commons, i.e. the atmosphere and the oceans – pertains to those areas to which economic rationality cannot be extended as easily as to the resources that are located within national boundaries. Nevertheless, Agenda 21 manages to conceive of the atmosphere and the oceans as a 'global resource' whose 'exploitation' needs to be regulated, the document says, through regional and global agreements.

THE MANAGEMENT OF HUMAN SETTLEMENTS

The fifth major theme deals with the management of human settlements, in particular in highly urbanized areas. Chapters here deal with the fundamental manageability of land-use problems, urban infrastructure, energy and transportation, the construction industry, and much else. These chapters speak to the heart of all kinds of engineers, urging them to build and develop in more energy- and resource-efficient ways — yet to build and develop nevertheless.

CHEMICALS AND THE MANAGEMENT OF WASTE

In the final theme — chemicals and the management of waste — Agenda 21 at least acknowledges that waste can be a problem, but states that chemicals are basically 'misused'. Agenda 21 does indeed include recommendations to reduce waste generation, to recycle waste materials into useful products, and to find safe methods of waste disposal. However, a detailed Greenpeace critique of the treatment of this theme in Agenda 21 concludes the following:[25]

Agenda 21:

- Does not recognize that there is no safe storage or disposal method for radioactive waste;
- Does not call for a ban on the dumping at sea of radioactive waste;
- Does not recognize that certain technologies like commercial reprocessing of radioactive substances produce more waste than others;
- Does not mention the nuclear contamination by military activity;
- Does not condemn the export of hazardous waste from industrial countries;
- Promotes the recycling of hazardous waste ignoring its toxic impact;
- Promotes voluntary rather than regulatory action for controlling pollution;
- Does not endorse guidelines for industry to use cleaner production techniques;
- Does not call for unlimited liability for trans-border nuclear pollution.

WHAT IS MISSING

Many themes were left out of Agenda 21. For example, it does not question Northern consumption patterns, nor the ones of Southern élites. All discussions on this topic were watered down considerably, by US negotiators in particular. Instead, Agenda 21 lays the burden of the problem on the population and individuals. On the other hand, the attempts to address the problems of overpopulation were shot down by the Vatican.

Needless to say that free trade and growth were left off the Agenda because they weren't even questioned. Like *Our Common Future* and *The Challenge to the South*, Agenda 21 does not question the enclosure and appropriation of common assets like seeds. It advocates free trade and implicitly endorses export. It pays lip service to communities by calling for global efforts that will encourage community solutions, instead of questioning the increasingly global economic system and institutions that are destroying the communities to begin with.

It does have chapters on the rights of women, children and youth, farmers and indigenous communities, a step forward over previous development declarations, though nothing of substance. For example, the recognition of indigenous communities is blunted by the fact that it addresses them as 'indigenous people', which legally means that they have individual rights. But the concept of 'indigenous peoples', which legally recognizes their claims as sovereign nations with rights of self-determination, was struck from every page of the document by the Brazilians and Canadians. Both these countries face considerable pressure from their indigenous communities who want to have more say in the use – or rather the destruction – of their own lands and resources.

The section on implementation in Agenda 21 is probably most symptomatic of its inherently flawed approach to the global crisis: Agenda 21 proposes simply more of the same old problem-solving techniques and technologies. That is, it proposes more information, more data, more training, more science and technology, especially technology transfer to the South, more money, and more and better international institutions. Though Agenda 21 mentions on numerous occasions the important role of women, indigenous people, local farmers, etc., their mobilization to implement Agenda 21 is always seen as part of a global management scheme orchestrated by international organizations.

Also, the UNCED secretariat estimated that it would cost about US$600 billion each year to implement Agenda 21, of which US$125 billion would be needed in aid for the South.[26]

In our view, more sustainable development aid – or in the language of Agenda 21, 'substantial flow of new and additional financial resources to developing countries' – will not solve any of today's global environment and development problems. Furthermore, it will probably exacerbate the existing ones, and create new problems. Also, if the major problem today is in the North, as Agenda 21 at times admits, then this money might be more effectively spent on fundamentally transforming Northern economies, instead of financing the export of Northern surplus products, most of which are environmentally and culturally destructive anyway. As a matter of fact, two of the most important chapters of Agenda 21 were finance and international institutions and the outcome of these will be discussed in Part IV.

The idea that technology is the solution – albeit a 'technology that does not further destroy the environment'[27] – is prevalent in Agenda 21. It is probably not exaggerating to say that technology is the biggest hope that emerges from UNCED in general and Agenda 21 in particular. Given the worldwide experience with technological progress over the past 100 years or so, the mythological belief in the miraculous emergence of fundamentally new, more efficient, cleaner, and environmentally safer technologies is probably, above all, wishful thinking. Needless to say that when Agenda 21 refers to technology it thinks first and foremost of high technology, which is fuelled by Western science. Biotechnology is probably the best example of the type of technology Agenda 21 is looking for. Not surprisingly, there is no mention of the socially and culturally disastrous effects that modern science and technology have had up to now. And only rarely do we find warnings about the potential environmental dangers of further technological progress. These critical remarks are generally made in regard to nuclear technology, but the risks are then immediately discounted as the price one has to pay for modernization. Risks, of course, can and must be managed.

The idea that more information will save the planet is equally disturbing, when knowledge about today's crisis is more than sufficient to take action. Why does anyone need more information, especially at the global management level at which UNCED wants to solve these problems, when the Brundtland report written by global managers had already drawn, in 1987, well informed and alarming conclusions? Not to mention the Club of Rome's 1972 report which provided sufficient information to take initial steps at least.[28]

Note that Agenda 21 does not talk of learning but of training, in particular training in technical knowledge – of which there is always more to know – as well as in all sorts of management skills, ranging from skills in 'sustainable management' to skills for global managers. By promoting this idea of a-sceptic knowledge and skills transmission – or, as they say, 'capacity-building' – Agenda 21 makes people believe, once more, that getting out of today's crisis is basically a technical problem. It would have been more honest and probably even more empowering to admit that we are in a dead end, and that collective – not individual – learning probably remains our only hope to find ways out of the present crisis.

CONCLUSION

In a recent interview the chief orchestrator of the UNCED exercise, Maurice Strong, admitted that:

there is no denying [that] the underlying conditions that have produced the civilizational crisis [that] the Earth Summit was designed to address did not change during the meeting in Rio. . . . The patterns of production and consumption that gave rise to so many of the global risks [sic!] we are dealing with are still in place.[29]

This is, of course, not surprising as the types of solutions proposed by UNCED – i.e. Western science, Western technology, Western information, Western training, Western money, and Western institutions – could not possibly have addressed the causes of the crisis, which happen to be Western as well. Rather, they were and still are the key fuelling forces of the process of industrial development, the very process that caused the crisis to begin with.

As we have tried to show in this first part, this flawed approach to solving the crisis can be traced back to the Brundtland and the South Commissions. The Brundtland Commission's contribution was to compromise environment and development through the use of the term sustainable development, while the South Commission's contribution was to begin to talk about the need for the South to band together and reassert that development was more important than environment. Malaysia, as it turned out, became the main voice of this argument.

Ex-President Nyerere's strong socialist background appears to have given the South Commission much more depth in its analysis of the past history of development than the Brundtland, but both believe in the same fundamental solution of stronger economic growth-oriented development. Like the Brundtland Commission, the South Commission lays a lot of blame at the door of population, but focuses much more on the inequality of the South's relationship with the North. Neither Commission challenges the development path of the North and both of them largely take Northern standards, means, and institutions as their goal – even though they pay lip service to their environmental and social unsustainability.

The South Commission's talk of popular participation in decision-making is somewhat stronger than in the Brundtland report, although there are few concrete suggestions on how to tackle this matter. Yet it ignores the idea of the commons and of community action. This is not really surprising, as both the South Commission and the Brundtland reports are basically government documents that are making suggestions on how governments can alter their existing policies while staying in power. Neither report attempts radically to deconstruct and analyse the problems. As a result, both Commissions reinforce the idea that the nation-state should have the power to solve the problems and support global management, as well as multinational and multilateral institutions.

Is it surprising that after foundation documents like *Our Common Future* and *The Challenge to the South* leaders on both sides – the environmentally concerned North and its apparent opponent, the poverty-stricken South – went to Rio with one message for the world: more growth, more trade, more aid, more science, more technology, and more management?

None of the treaties and agreements signed in Rio tackles any of the major causes of environmental problems, such as the pressure placed on the planet by Northern consumption or unsustainable patterns of development in the South. The problems of free trade, militarization, and mega-polluters like some multinational companies have been dropped completely. The agreements on stemming the most obvious symptoms of environmental problems like global warming, desertification, and loss of species and forest cover have no real targets. It is hard to imagine that they will make a difference.

Part II

NON-GOVERNMENTAL
ORGANIZATIONS

When Maurice Strong was given the job of Secretary-General of the Earth Summit, he announced an ambitious plan to involve millions of people in the UNCED process. He even promised NGOs access to the negotiations. The UNCED secretariat made a valiant effort to try and bring in previously unheard voices by setting up a special NGO liaison unit to assist NGOs from all over the world to come to the PrepComs and lobby the government delegates on whatever aspect of the agreements they thought was important. Other organizations such as the Center for Our Common Future embarked on a similar effort by organizing what they called the 'independent sector', i.e. all people, groups, and organizations that are not officially linked to the governments. And many other NGOs started to form federations so as to become more efficient in influencing the UNCED process.

At the beginning of the UNCED process, the results were rather discouraging. A mere thirty NGOs turned up in Nairobi for the first PrepCom meeting in August 1991. And the governments who were used to talking behind closed doors were very reluctant to let them into the talks. But as the negotiations got under way, the numbers swelled and the governments relented. In New York at the final PrepCom meeting in March 1992 there were about 1,000 accredited NGO participants. In Rio 1,420 NGOs registered with the secretariat. Ostensibly for security and space reasons, most of them were not allowed into the government negotiations. Instead, the Brazilian government worked with a number of NGOs to help them (and many other NGOs which were not accredited with the secretariat) to hold meetings among themselves in some specially constructed tents in Rio de Janeiro's Flamengo Park, 40 kilometres away from the official discussions. The media, however, focused on the governments and their negotiations and treated the NGOs mainly as a joke.

Despite that, many NGOs came away pleased with their own efforts, feeling that they had contributed to saving the planet. But many others also came away feeling frustrated by the Summit and the UNCED process itself. Many criticized the governments and said that the two years of negotiations had achieved very little. In order better to understand the NGOs' judgement on Rio, let us therefore analyse how NGOs became part of the UNCED process (chapter 5) and assess what they finally achieved (chapter 6). However, before we can do that, it is important to understand the diversity of the Green movement, as this explains, in part, the movement's relationship to and involvement in the UNCED process. This is what we turn to in the next chapter.

4

TELLING 'GREENS' APART

If we want to understand who went to the PrepComs and to Rio, who did not, and who achieved what, it is necessary to get a better sense of what the Green movement is, where it comes from, and where it is heading. As mentioned earlier, over 1,400 NGOs were accredited at the Earth Summit and about 30,000 people — concerned citizens, activists, and NGO representatives — showed up at the Earth Summit or during the Preparatory Meetings. Yet many others did not go to Rio, and among the ones who did very few turned out to be effective.

Although historically the movement has largely been in opposition to the 'system', to the 'establishment', and to governments, as well as to business and industry, over the years parts of the movement have become bureaucratized and part of the establishment themselves. Some groups were transformed into political parties, others into NGOs, and still others into lobbies. New groups formed and others disappeared. Substantial changes are also observable in terms of topics and issues: these have substantially changed over time, as new topics have emerged and old ones become outdated. As a result, the Green movement, at the beginning of the 1990s, is quite powerful yet diverse and at times fragmented. But the Green movement, like the establishment, was taken by surprise by the new trend towards global ecology in the second half of the 1980s: it had not anticipated that trend, much less promoted it, and was faced like all other major societal agents with the need to redefine itself in the light of this new trend. As for all other agents, the UNCED process offered the Green movement a place and a chance to do this.

Yet by and large the Green movement has failed to achieve this goal: it did not emerge from Rio stronger, but weaker. As a result, it is more fragmented and more disoriented than before. As we show later on, this is partly the result of deliberate efforts not so much to destroy the movement directly, but to feed it into the UNCED process. But if the Green movement comes out of Rio weaker, this is also because it was already quite fragmented. In many ways, Rio has simply exacerbated this fragmentation. Therefore, if we want to understand the role the Green movement did – or rather did not – play in Rio, it is necessary to assess the movement's main trends over the past twenty years or so.

When trying to put some order into the Green movement, there are some necessary distinctions that need to be made. First, one needs to distinguish between the Green movement in the North and the one in the South. Indeed, though in Rio many activists celebrated the so-called 'same-boat-mentality', global awareness, and South–North unity, one should not forget that the movements in the North and in the South have evolved separately around quite different issues. And though the idea of being in the same boat is a tempting one given the new global ecology, the strategic agendas of the movements in the North and in the South are not the same. And this, at least in part, explains why this global NGO alliance, expected and hoped for by many in Rio, never really materialized.

THE GREEN MOVEMENT IN THE NORTH

The Green movement in the North, like the one in the South, is far from being homogeneous.[1] Similarly, there was hardly a Green movement of any significance in Eastern Europe before Gorbachev came to power in 1986. Though there were local protests in Eastern Europe, especially against the state-controlled nuclear industry, there was no Green movement comparable to the one in Western Europe or Northern America. But during *perestroika* and *glasnost* in the Soviet Union, and with the prospect of independence for the Eastern European countries and Soviet republics, the Green movement in the East grew exponentially.

Not surprisingly, in a highly politicized society and at a highly political moment, the Green movement in the East was first and foremost a political movement with political, i.e. national, agendas. Be it in Hungary, Poland, Czechoslovakia, or Estonia, the Green movements turned rapidly into Green parties, which in turn quickly acquired a share of national political power. But once this had taken place, the Green movement declined. In retrospect, it turns out that the Green movement in the East was instrumental in the transition of the Eastern European countries to a market economy. Yet, despite the enormous ecological problems facing the East, in the 1990s the Green movement has substantially lost momentum and declined. Except for some localized protests – for example against the damming of the Danube in Hungary, or against nuclear energy in Russia and Ukraine – the Green movement in the East has more or less become insignificant. In Rio, the Eastern Green movement was basically absent.

As for the Green movement in the West, it is necessary, to our mind, to distinguish between the movement in Western Europe and the one in the United States. The basic difference is in terms of their relationship to politics and the political system. If in Western Europe we are dealing with a political Green movement, in the United States we are basically dealing with environmental lobbies. This is not to say that environmental lobbying organizations – such as the WWF, IUCN, and all kinds of others – were not important in Western Europe. However, in the late 1970s these organizations were rapidly bypassed by political Green groups, in particular by the anti-nuclear movement.

It is out of these groups that the various Western European Green parties emerged at the end of the 1970s. This is particularly the case in West Germany, the Netherlands, Belgium, Switzerland, Austria, all four Scandinavian countries, and Italy. The Green parties in France, Spain, Portugal, and Greece took longer to emerge, but can also be traced back to the political ecology movements of the 1970s. All Green parties generally picked up votes in the early 1980s, but declined again towards the end of the 1980s and in particular since the beginning of the 1990s. But in the mean time the success of the green parties had substantially weakened most other environmental agents in Western Europe, except perhaps for the older nature protection organizations. Only a limited number of agents from the political ecology movement continued to thrive. We think here in particular of Friends of the Earth and Greenpeace. Though on an ideological level they both arose out of the political ecological movement of the beginning of the 1970s, they have managed to survive the decline of the Green parties. However, of late they have also had to face a loss in public support and financing.

The Green movement in the United States is just as much the product of the 1960s and the early 1970s as is the movement in Western Europe.[2] But if in Western Europe the Green movement basically ruined itself by competing for power with the traditional political parties, in the USA the Green movement was, by contrast, ruined because it was used by conservationist environmental lobbying organizations. Indeed, the US political system and the perceptions of US citizens is such that lobbying is believed to make all the political difference. With three exceptions, all major environmental lobbying organizations in the USA originated around issues of nature protection and environmental conservation. Today, these organizations are known as 'the Big 10': the Sierra Club (founded 1892), the National Audubon Society (1905), the National Parks and Conservation Association (1919), the Izaak Walton League (1922), the Wilderness Society (1935), the National Wildlife Federation (1936), the Defenders of Wildlife (1947), the Environmental Defense Fund (1967), Friends of the Earth (1969), and the Natural Resources Defense Council (1970).

These ten are among the wealthiest environmental organizations in the United States and probably in the world. The problem is that they are basically mainstream and effectively monopolize public support for environmental issues in the United States. Though they have added some elements of pollution control to their conservationist agenda, they limit themselves to lobbying the political system by calling for more efficient environmental management. Not surprisingly, the chief executives of the Big 10 are mainly lawyers, generally earning as much as their counterparts in business. One of the authors once went to what he thought was a World Bank meeting with NGOs and discovered only half an hour later that he was in the wrong meeting. It was in fact a caucus of the Big 10. But it was easy to see how one could be fooled by looking at the table and seeing a dozen white men dressed in sharp business suits. It is also important to note that this is the way the Big 10 choose to work, i.e. they feel they can be most effective, or convincing, when meeting the bankers on their own terms and putting them at ease.

Of course, environmental activism in the United States is not limited to the Big 10. But, unlike in Western Europe, the political Greens in the USA have never really taken off. The US Green party, for example, is virtually non-existent. Only a few sizeable organizations such as the Earth Island Institute or to a certain extent Earth First! can be considered political Greens. On the other hand, the US Green movement differs from the one in Europe in that it has quite a strong ecological, as well as a New Age, strand. Let us now look at the various trends within the Green movement of the North in order better to understand its fragmentation.

THE MAIN TRENDS OF THE GREEN MOVEMENT IN THE NORTH

We can observe in the evolution of the Green movement in the North over the past 20 years four major trends, all of which came to play a role in Rio. These are the transformation of conservationist ecology into global environmental management, the erosion of political ecology, the radicalization of some parts of the Green movement, and the trend towards New Age environmentalism.

FROM CONSERVATIONIST ECOLOGY TO GLOBAL ENVIRONMENTAL MANAGEMENT

The most significant trend is without doubt the transformation of con-servationist ecology into global environmental management. Indeed, conser-vationist environmentalists such as most of the Big 10, as well as WWF and IUCN, have always been interested in scientific environmental management. Not surprisingly, they closely collaborate with governments in order to promote environmental management policies. Global ecology – which is above all a product of a new global science and thus contains a heavy technocratic bias – has rapidly been embraced by these conservationist organizations. They saw in global ecology the logical next step of their endeavours: from a national level, environmental management quite logically must now move to a global level. Interestingly, IUCN had already coined the term 'sustainable develop-ment' in 1980, though the term still had a slightly different meaning.[3]

Indeed, for the new global managers of the Brundtland Commission and other UN agencies, the conservationist environmentalists were probably the most natural allies, provided however they made some revisions in their conservationist philosophies – which they gladly did. In 1986, for example, the big US environmental organizations – most of the Big 10 – issued a joint statement entitled 'An Environmental Agenda for the Future' which described environmental pollution as a technological challenge, rather than an economic, political, and social issue.[4] It did not criticize US dependency on petro-chemicals, made no mention of nuclear energy, and offered no strong recommendations in support of increased reliance on renewable energy, organic farming, sustainable-yield logging, or mass transit.

Two years later, WWF literature started to blame the poor for being the 'most direct threat to wildlife and wildlands'.[5] With poverty being identified as an environmental problem and technology as its solution, most con-servationist environmentalists have put themselves, since the second half of the 1980s, in line with the ideology of the Brundtland Commission and the UNCED on sustainable development. They were now ready to be admitted to the club of global environmental managers. With this came an even closer relationship with business and industry, as business also was under pressure to become green and was thus looking out for partners in the environmental movement. WWF, for example, received US$50,000 each from oil companies Chevron and Exxon in 1991. The National Wildlife Federation conducts enviro-seminars for corporate executives from such chemical giants as du Pont and Monsanto for a US$10,000 membership fee in their Corporate Conserva-tion Council programme. The Audubon Society meanwhile sold Mobil Oil the rights to drill for oil under its Baker bird sanctuary in Michigan, garnering US$400,000 a year from this venture.[6]

In short, this first trend led the big conservationist organizations – in particular the US Big 10, the WWF, IUCN, and the World Resources Institute – to become part and parcel of the global environmental management establishment. They are now basically promoting the same global environmen-tal management that UNCED was striving for. In Rio these conservationist organizations had substantial lobbying power and access to the negotiations. But unfortunately, by the time they had achieved such access, they had become so mainstream that their input can hardly be detected in the Rio documents.

THE EROSION OF POLITICAL ECOLOGY

The second trend to be noted in the Rio process is the erosion of political ecology in the North. As we have seen, political ecologists have mainly focused on national politics. Though they were highly critical of national politics, they believed that national policies could be changed by their participation in the national political system. We have also observed that political ecology was already declining in the late 1980s, but it was really the emergence of global ecology that gave political ecology the final blow. Preoccupied by national and regional eco-political issues, political ecologists totally missed the trend towards global ecology. As a matter of fact, they were taken by surprise. Even today, the remaining political ecological groups struggle to adapt to this trend. This is, for example, the case of Friends of the Earth which tries to redefine itself while downsizing. Friends of the Earth was not very noticeable in Rio, and most political ecological groups were absent. The only exception to this was Greenpeace which — with the largest membership of any environmental group in the world (2 million) and a budget bigger than that of the United Nations Environment Programme (approximately US$150 million) — continues the tactics of confrontation while keeping up its lobbying. The tactics have earned it criticism from both sides: the Big 10 see Greenpeace as a bit of an outlaw, while the more radical and grassroots environmentalists describe it as too corporate.

THE RADICALIZATION OF PARTS OF THE GREEN MOVEMENT

The third trend, therefore, is the radicalization of the Green movement in the North and perhaps also in the South. Given the erosion of political ecology, this trend highlights a new polarization, i.e. a trend towards protest and, to a certain extent, eco-fundamentalism. This trend can be associated with the deep ecology movement, for which groups like Earth First!, Wild Earth, the Sea Shepherds and others have become representatives.

Unlike Greenpeace, which confronts but never proceeds beyond that, the deep ecologists do not exclude property damage. Former Greenpeace (Canada) founder, later expelled from that organization for his radical tactics and now leader of the Sea Shepherds, Paul Watson, said: 'Pardon me for my old-fashioned ways, I believe that respect for life takes precedence over respect for property which is used to take life'.[7] Contrary to popular belief, deep ecologists do not condone physical injury to human beings, although their tactics have run the risk of doing so.

Yet their tactics have often been surprisingly effective. The Sea Shepherds closed down the Icelandic whaling industry singlehandedly one cold November night in 1986 by the simple expedient of sinking two of its four ships and destroying the refrigeration system of its whale processing plant. Others have gone further. Eco-saboteurs in Canada blew up a US$4.5 million hydro-electric substation on Vancouver Island in 1982. In Thailand they burnt down a tantalum plant in 1986 causing damage estimated at US$45 million. Lapps in Norway blew up a bridge leading to a dam that had flooded their lands.

Then further out from even the deep ecologists are people who do accept physical injury or death as punishment. The motives of these are probably closer to revenge. For example, Primea Linea, an Italian group, claimed responsibility for machine gunning Enrico Paoletti, an executive of a Hoffmann-LaRoche subsidiary, who was in charge of the chemical plant in Seveso, Italy, that exploded in 1976 to release a dioxin cloud. Primea Linea claimed that it was delivering just punishment for his deeds.

To a certain extent, this trend towards deep ecology is simply the other face of the aforementioned trend towards global management: both share a quite unsophisticated analysis of the socio-political dimensions of today's global ecological crisis and blame the 'humans', or as Brundtland says 'human development', for today's crisis. However, unlike the global managers who made Rio their event, deep ecologists were hardly present in Rio, preoccupied as they were by fighting the concrete everyday local destruction of the environment. At times, deep ecologists spoke in Rio through the youth representatives and at times through the representatives of indigenous peoples.

THE TREND TOWARDS NEW AGE ENVIRONMENTALISM

The issues of indigenous peoples in particular have been subsumed by a fourth trend which became particularly visible in Rio: the trend towards New Age environmentalism. The New Age phenomenon has been rampant in Western societies, especially in the United States and Canada, since the middle of the 1980s. Though New Age ideas can be traced back to the hippie movement of the 1960s, it is the new globalization of the 1980s combined with rapidly growing individualism that have made the New Age a significant societal phenomenon in the Western hemisphere. Rio made New Age globally prominent, and for the first time exposed the whole world, especially the South, to this typically Northern phenomenon. Maurice Strong, the chief orchestrator of Rio, is himself a quite typical representative and vocal advocate of the New Age.[8] At Rio, the New Age took the form of a celebration of global environmental awareness mixed with a touch of spiritualism.

Global management and the New Age, the two most visible trends in Rio, are far from being contradictory. Rather, they reinforce each other. On the one hand, the idea that all individuals are now connected because they all share a common global environmental awareness quite logically leads to some sort of global management. On the other hand, global management is probably in need of some sort of 'philosophical' framework that would give it the moral and ethical dimensions necessary for it to be legitimized by the people. Not to mention the fact that many of the global managers are themselves members of the New Age church. In our view, the biggest problem with the New Age religion is that it is a-political, a-sociological, a-cultural, and a-rational. Presenting, as UNCED did, the global environmental crisis as being the result of a lack of global environmental awareness was, of course, inadequate. But it fulfilled a particular function: not only could the global managers display such awareness, they could also argue convincingly that they already occupied the key positions from where such an awareness would actually make a difference.

In short, we can say that over the past 20 years the Green movement in the North has undergone a substantial transformation. And this process has certainly been accelerated by UNCED. First, conservationist environmentalism has been promoted to the level of global management, whereas the political ecology movement has been further eroded. As a reaction, we observe further radicalization among some environmental groups, though this did not have any significant impact on Rio. Finally, Rio significantly helped New Age environmentalism to come forward, legitimizing New Age. This New Age environmentalism has given global management the 'philosophical' backing it was lacking until now.

THE GREEN MOVEMENT IN THE SOUTH

In order to understand the participation of Southern NGOs in the UNCED process, one has to recall that in the South the evolution of civil society in general and the Green movement in particular has been quite different to that in the North. One has to start with the colonial past of most developing countries and their striving for economic and political development. The first development decade – the 1960s – was thus characterized by optimism, as well as by the idea that Northern style development was achievable and desirable. Local communities and local organizations were above all seen as impediments to development. Consequently, the so-called 'NGOs' – international and national ones – were basically charitable organizations. They were often religiously inspired, generally top-down, and mainly organized by the North.

But as this top-down or one-way development strategy did not seem very effective, a new type of NGO, so-called 'second generation NGO', emerged.[9] As a matter of fact, the various groups working for development became more independent of the North and started to work with local agents operating on a smaller scale. They began to strive for self-reliance and development from the bottom up. The emergence of these second generation NGOs in the South cannot, of course, be separated from the world's economic crisis in the 1970s and from a certain disillusionment with the role of the state and the international economy in Third World development. NGOs, during the 1970s, were therefore essentially about popular participation. At the same time they increasingly became seen by Northern aid agencies as a channel for distributing aid more effectively.

During the 1970s and within this second generation NGO framework, a series of environmentally oriented NGOs emerged in the South. These included the Environmental Liaison Center International (ELCI) created in 1974, Environment and Development Action in the Third World (ENDA) in 1976, Sahabat Alam Malaysia in 1977, and the Green Belt Movement in Kenya in the same year. These and many other environmental NGOs in the South already operated at that time within an environmental and developmental framework. At least at local and regional levels, environmental protection and restoration could not be separated from development. Indeed, environmental restoration such as tree planting was considered to be an integral part of participatory and self-reliant development.

To be sure, none of these NGOs opposed the idea of development, thus still buying into the Northern development ideology. But the fact that it was said to be people's development – as opposed to governments' or multinationals' development – made this endeavour, in the eyes of many, sufficiently different, so as to coin the term 'another development'. Ideologically, the idea of another development in the South was very close to, and often inspired by, the political ecology movement in the North, in particular Western Europe.

As the second development decade ended, self-reliant participatory development in the South can show some small successes. However, in the South overall the situation worsened: Third World countries' debt increased steadily, their environment was more degraded, poverty and malnutrition grew, and many countries' prospects for future development looked rather bleak. In the third development decade, the 1980s, many in the NGO movement therefore come to recognize that self-reliant participatory development has certain limitations, mainly because the NGOs' 'scope of attention is limited to individual villages and neighborhoods, and to specific local groups the NGO is assisting'.[10] Second generation NGOs were unable to deal successfully with the real causes of underdevelopment and environmental decline in the South, which mainly pertain to the globalization of the economy. And this led to a split of the Green movement in the South during the third development decade, i.e. the 1980s.

On the one hand we can observe the transformation of the second generation NGOs into 'third generation NGOs'.[11] Indeed, local initiatives in the South now become increasingly seen within a national and international environmental and developmental framework. As Korten puts it, 'self-reliant village development initiatives are likely to be sustained only to the extent that local public and private institutions are linked into a supportive national development system'.[12] Having withdrawn from and criticized governmental initiatives, NGOs in the 1980s, according to Korten, were again seeking collaboration with governments and international agencies. International agencies in turn, operating under a new awareness of global ecology, were trying to integrate NGOs into this global environmental and developmental framework. It is from this background that many grassroots and local NGOs organized themselves during the 1980s into coalitions, many of them with a strong environmental component. Since the beginning of the 1980s various NGO coalitions have emerged, such as APPEN (Asia Pacific People's Environmental Network) in 1983, ANEN (African NGOs Environmental Network) in 1986, or ANGOC, the Asian NGO Coalition. Today, it seems, Southern NGOs are seeking more and more to organize themselves regionally, nationally, and even internationally.

As a result, many of these Southern NGO coalitions have become quite powerful agents, often bypassing national governments and negotiating directly with international donor agencies. Their framework remains one of participatory development, generally with a particular focus on local environmental resources management. Quite logically, many of these Southern NGO coalitions went to Rio. And even some local NGOs did. At least on paper the Rio framework was totally in line with their own approach to environment and development. However, in practice, concrete participation in the UNCED process turned out to be quite difficult for them. One of their major problems was the lack of organization. As a result, many Northern conservationist NGOs functioned as a voice of Southern NGOs.

But there is also, on the other hand, another trend detectable during this third development decade in the South. We refer here to environmental protest movements, somewhat similar to the political ecology movements of the 1970s in the North. In the South, such groups criticize and protest against Northern development schemes, promoted by such international agencies as FAO, UNDP, or the World Bank, and implemented with the help of national and local élites. These groups oppose, for example, large dams, modern industrial agriculture and reforestation schemes, transmigration programmes, deforestation, and other Northern-inspired, massive development efforts. They criticize, along the lines of a political ecology approach, Northern science, technology, and more generally industrial practices put forth by transnational corporations, their own national governments, northern governments, and international development agencies.

In 1989 the Kayapo Indians, for example, protested against the building of a US$10 billion World Bank dam near Altamira in the state of Amazonia. Six thousand locals and Kayapo joined a five-day rally with British rock singer Sting. Dressed in full battle gear they waved clubs and spears at the engineers. This was not their first fight, as they had held gold miners hostage four years ago and camped out in the parliamentary buildings of Brasilia for days campaigning against nuclear waste. The World Bank caved in and the dam, which would have displaced 70,000 people, was cancelled. In India, Sunderlal Bahaguna helped start the Chipko (tree huggers) movement in 1973 to protect forests from local commercial development and prevent the landslides that accompanied deforestation. The movement later led to tree planting and has grown to include protests against the environmentally unsound Tehri dam and limestone quarries.[13]

The Third World Network, founded in 1985, and its journal *Third World Resurgence* have become a voice for many of these protest movements of the South. And the Third World Network, together with other like-minded organizations protesting against the destructive effects of Northern-type development in the South, was highly prominent in Rio. Unlike the declining political ecology movement in the North, this Southern political ecology movement is thriving. But like the political ecology movement in the North, its relationship to national politics is highly ambiguous, as we show in the case of Rio: though the Third World Network, for example, sees Northern inspired industrial development as a problem for the South, it nevertheless believes that political control in the Southern countries is the answer to the problem.

Also, because of the visibility of such environmental protests in the South and the negative consequences this has for obtaining further financial support from the North, this political ecology movement in the South has become a threat to some Southern élites as well. Through the Rio process, however, they managed to channel this protest into a North–South framework, thus using the Southern political ecology movement in order to put additional pressure on the North. As a matter of fact, the Third World Network, for example, has actually provided many arguments to the Southern élites, and helped them to reposition themselves in the light of the new challenge of global ecology. Building on this political analysis of environmental problems as put forth, for example, by the Third World Network in Malaysia or the Center for Science and the Environment in India, it is no longer industrial development *per se* which is considered destructive of the environment. Rather it is the fact that development remains controlled by the North instead of the South. The weakness of this argument, of course, stems from the fact that it mixes together Southern peoples and Southern élites.

RIO AND THE VARIOUS SHADES OF GREEN

This historical look at the Green movement shows that, by the time the UNCED process took place, this movement was already quite fragmented. This fragmentation was due, in part, to the fact that the global ecology has created a new situation to which the Green movement – along with many other societal agents, including business and industry – was and still is trying to adjust. Therefore, since the 1980s, the Green movement has seemed to be quite disoriented. The least disoriented groups, i.e. the ones with the most coherent intellectual framework, turned out to be the most efficient in Rio. This is particularly true of the big conservationist organizations – the Big 10, WWF, IUCN, and WRI – as well as the Third World Network. The other three factions of the Green movement – i.e. the New Age Greens, the Northern political Greens, and the Southern participatory (environment and) development coalitions – turned out to be more disoriented, confused, and not surprisingly, quite disorganized. As we will see, various efforts were made to organize these fragmented groups, in particular by the UNCED secretariat, the Center for Our Common Future, and environmental groups themselves. The Northern political Greens and some Southern NGO coalitions tried to team up in an effort to have their own 'Social Movement Summit', whereas the US Citizens Network and the Canadian Participatory Committee for UNCED were trying to organize the New Age environmentalists. Greenpeace remained outside all these coalition-building and organizing efforts but was prominent in the UNCED process. Finally, deep ecologists and social ecologists stayed away from Rio, considering from the very beginning that UNCED was going to be a 'débâcle'.[14]

It is still too early to assess what effects this débâcle will have on the Green movement. Yet it is likely that those organizations that were riding the wave of Rio – the conservationist environmentalists and the Southern political ecologists – will sooner or later have to deal with this 'débâcle' and the effects it will have on them. This is not to say that the other factions of the Green movement will be better off in the long run, considering in particular the fact that they have not clarified their role in the age of global ecology either.

5

FEEDING THE PEOPLES INTO THE GREEN MACHINE

The Green movement was confronted by the UNCED process, as we have seen, at a particularly crucial moment in its own evolution, when it was fragmented and needed to redefine itself in the light of the new global ecology. Would it be capable of seizing UNCED as an opportunity and redefining itself? This was the question before Rio. Today the questions in our mind are rather: what did UNCED do to the Green movement, to its various organizations, groups, and NGOs? And what, in turn, did the movement 'do' to UNCED? To what extent did it influence the UNCED process and the Rio conference? Those questions are examined in chapter 6. First, in this chapter, we look at how the Green movement got fed into the UNCED process. In the first section we examine how the UNCED officials conceptualized the Green movement's role. We then look at the various organizations that sprang up in order to feed NGOs into the UNCED process. Finally, we assess the role the Green movement played in UNCED, and show that the movement basically became coopted into the Rio process.

THE OFFICIAL VISION OF NGO PARTICIPATION IN UNCED

There is no doubt that the Green movement was actually taken by surprise by the Rio process, in the same way that it was taken by surprise by global ecology. Consequently, the movement had no strategy on how to respond. Maurice Strong, however, did have a vision. It is important to understand his vision, as it came to drive the way NGOs related to and were fed into the UNCED process. It is a vision which, once applied, turned out to favour certain NGOs, groups or organizations at the expense of others. On the other hand, it is also a vision which ideologically is quite attractive. Many people have embraced it quite enthusiastically. So did the Green movement, but by doing so it bought into an approach which ultimately weakened it. In order to understand this vision we have to backtrack a little and trace its origins.

To a large extent, the UNCED model of NGO or civil society participation is the international establishment's answer to the global crisis. The establishment's reference point for this crisis has been the Cold War. With the emergence of global ecology – and in particular the ozone hole at the beginning of the 1980s – the international establishment started to see the global environment as yet another way of overcoming the East–West divide and fostering dialogue and cooperation among heads of state. We have many reasons to believe that at the start of the Brundtland Commission's work, the goal was not so much to solve global environmental problems as to create opportunities for dialogue. Some of this desire for dialogue might of course have been driven by commercial interests, seeking to extend business to the other side of the Iron Curtain.

In the mind of the international establishment, the global ecological crisis thus became re-framed in terms of the threat of nuclear weapons. What is more, environmental degradation was considered to exacerbate that threat. Therefore it reinforced the idea promoted by the establishment, that we are all in the same boat, i.e. what we call the 'same-boat-ideology'. This ideology says that global environmental degradation – like nuclear weapons before – is a threat to all inhabitants of this planet alike. We therefore are all sitting in the same boat and have no choice but to engage in dialogue and cooperate, as we will either win or lose together. The responses to the global crisis as implied by the 'same-boat-ideology' are, therefore, (1) dialogue among enlightened individuals, (2) global environmental awareness raising and corresponding ethics, and (3) planetary stewardship. All three responses are actually rooted in the Brundtland report's failure to identify the real causes of today's global ecological crisis, and can be related to the international establishment's obsession with Cold War problems. As a result, one is left with the impression that the only reason why the global ecological crisis exists to begin with is because of a lack of dialogue between the citizens and their leaders and between the leaders themselves. Note that this same-boat-ideology is a significant component of the above mentioned New Age environmentalism.

But if dialogue was, perhaps, an efficient means of ending the Cold War, it has become, within the UNCED process, a goal in itself. As such, the obsession with establishing dialogue has diverted attention from the real issues, perpetuated business as usual, and contributed to coopting and weakening the Green movement. This same-boat-ideology attributes a key role to NGOs who are said to be partners in dialogue, as well as multiplicands of environmental awareness and carriers of planetary responsibility.

Consequently, the two-year-long UNCED preparatory process was essentially designed to achieve two things, i.e. to build a so-called UNCED constituency by getting NGOs and, even more so, NGO coalitions to support UNCED publicly, and to identify some NGO or independent sectors' leaders as associates in global management. For these two purposes the UNCED secretariat created a special NGO-liaison office, whereas the Center for our Common Future came up, in June 1990, with an International Facilitating Committee (IFC) to help NGOs become part of UNCED. Moreover, many NGOs themselves made efforts to feed into this UNCED process. All this, plus the accreditation procedure which made it easy for all interested NGOs to become part of the UNCED process, should have helped to build this strong UNCED constituency and select potential working partners in global management. Let us therefore look at what UNCED itself put in place, how the Center for our Common Future tried to feed NGOs into UNCED, and how NGOs organized themselves in order to become part of UNCED.

NGO ACCREDITATION

Despite the Brundtland Commission's vision of a global dialogue among all partners who sit in the same boat – that is, basically all inhabitants of this planet – and despite the desire of Maurice Strong and his secretariat to get NGOs involved in the Earth Summit, persuading the governments to do so was a difficult task. General Assembly Resolution 44/228 of December 1989 which sets the UNCED process into motion requests 'relevant nongovernmental organizations in consultative status with the Economic and Social Council to contribute to the Conference as appropriate'.[1] In a preparatory document by the Secretary-General at the UNCED organizational session in New York in

March 1990, it was stated that the community of non-governmental organizations could:

enrich and enhance the deliberations of the Conference and its preparatory process through its contributions and serve as an important channel to disseminate its results, as well as to promote the integration of environment and development policies at the national and international levels, and [that] it is therefore important that non-governmental organizations contribute effectively to the success of the conference and its preparatory process.[2]

The Secretary-General, Maurice Strong, was therefore invited by the Preparatory Committee to propose arrangements for NGO participation at the first PrepCom in Nairobi in August 1991.

Besides the idea of a dialogue among partners in global management, which had been put into practice for the first time in the Brundtland Commission's hearings, Maurice Strong used yet another argument in order to sell NGO and independent sector participation to governments. This was the idea that NGOs would contribute information to UNCED on the one hand and help disseminate its outcomes on the other, while the governments would remain in charge of the whole process. Therefore, getting NGOs and independent sectors to become involved in the UNCED process would not only create a dialogue among all partners in global management, it would also legitimize all governments involved in the UNCED process. Needless to say, many governments had some problems with this argument.

The first time this NGO participation model was tried out was in a regional conference preparing for UNCED, i.e. the follow-up conference on Environment and Development for the European region held in May 1990 in Bergen. This so-called Bergen conference took place at a ministerial level, but there was a planned attempt to involve what was called 'the independent sector' in the discussions with the ministers. By 'independent sector' one must understand organizations which are supposedly independent from government, such as industry, trade unions, the scientific community, youth, and NGOs. And the *Brundtland Bulletin* concluded optimistically: 'The Bergen process of consensus-seeking between independent and official channels . . . seems to become the model for the '92 process'.[3]

As a result of the Bergen meeting, Secretary-General Maurice Strong met with the representatives from the independent sector, including the board of CONGO and other non-governmental organizations, and stressed his support for the principle of broad representation and participation. The precedent set by Bergen meant that more or less any relevant sector in society — and not just environment and development NGOs — could and should be involved in the UNCED process. Moreover, the various NGOs were asked to organize themselves into coalitions, so that ultimately the various sectors would end up speaking with one voice.

After the Bergen precedent backed by Maurice Strong's vision, the stage was set for the debate about NGO participation at the first PrepCom in Nairobi. To recall, the Economic and Social Council to the General Assembly of the UN (ECOSOC) allows certain accredited NGOs to attend its meetings in New York and some are even allowed to suggest topics for discussion. There are upwards of 900 NGOs which have this 'consultative' status and they form a caucus called CONGO. As mentioned above, UN Resolution 44/228 called for their inclusion in the UNCED process, but said nothing about other NGOs. The UNCED secretariat, however, had its own interpretation.

At the first PrepCom meeting in Nairobi in August 1990, government delegates discussed this issue for the first time. Secretary-General Strong opened the session by noting that, though desirable, the Bergen formula for NGO participation would not be realistic or applicable, given the fact that the number of both governments and non-governmental organizations would be much greater in Brazil than in Bergen. He did recommend, nevertheless, that the Bergen 'principles' be applied. In the ensuing debate about NGO access, Tunisia and Mauritania objected to NGO access. Bolivia, on behalf of G-77, was in favour of NGO access but only for those with consultative status at ECOSOC. 'After several days of intense discussion', Dawkins reports, 'the Preparatory Committee acknowledged that the effective contribution of non-governmental organizations was in its interest, but approved a far narrower role than the one encouraged by the Bergen Ministers'.[4]

The final outcome of the Preparatory Committee's deliberations with regard to NGO participation can be summarized in the three following points:[5]

1. Non-governmental organizations shall not have any negotiating role in the work of the Preparatory Committee.
2. Relevant non-governmental organizations may, at their own expense, make written presentations in the preparatory process.

3. Relevant non-governmental organizations in consultative status with ECOSOC may be given an opportunity to briefly address plenary meetings of the Preparatory Committee and meetings of the working groups. Other relevant non-governmental organizations may also ask to speak briefly in such meetings. However, this would be at the discretion of the Chairman and with the consent of the Preparatory Committee or the Working Group.

Overall it appears, therefore, that NGO accreditation to the UNCED process is different from standard UN practice only insomuch as NGOs without consultative status with ECOSOC are granted the same rights as the ones with such consultative status. And this was actually the minimum condition needed in order to get NGOs to become part of the UNCED process. NGOs and all other sectors seemed more or less happy with this formula. As a result, they asked to be accredited to UNCED in great numbers, a process which the secretariat handled. Basically, anyone who wanted to be accredited could do so. And according to our information, only four out of 1,420 NGOs were refused accreditation with the UNCED secretariat over the entire two-year process. If the UNCED secretariat was accrediting NGOs, the Center for Our Common Future took an active role in feeding them into the UNCED process. But in order to understand the exact role the Center played, we have to understand its origins.

THE CENTER FOR OUR COMMON FUTURE

After the Brundtland report had been submitted to the UN General Assembly in October 1987, and the Brundtland Commission had been officially dissolved in December 1987, the question arose as to what to do next. As Warren Lindner, then Secretary of the Brundtland Commission, reports:

Ultimately it was decided that I would establish a charitable foundation called the Center for Our Common Future, whose sole agenda would be to further the messages contained in the report and broaden the understanding, debate, dialogue and analysis around the concept of sustainable development. The Center would move that debate into as many sectors of society and as many countries as possible.[6]

In other words, the Center for Our Common Future was conceived as a public relations agency, making publicity for the Brundtland report. Thus, the Center was established in April 1988 with voluntary funds and located at the same site as the Commission had been. Most of the staff transferred to the Center.

In the first phase, the role of the Center was basically to spread the message of sustainable development as well as the Brundtland report itself. During that time, the Brundtland report was translated into over twenty languages, accompanied by a video as well as educational and promotional materials. Moreover, the Center built up a network of over 160 so-called 'partners of the Center' in about seventy countries worldwide. These partners included not only intergovernmental organizations, environmental and developmental NGOs, media representatives, youth, women, and financial organizations, but also trade unions and professional organizations. On the environmental side, the working partners of the Center include IUCN, WWF, WRI, the European Environmental Bureau (EEB), and many other establishment NGOs. Working partners, as well as all other people and other organizations who happen to have been in contact with the Center are modestly called the 'Brundtland constituency'. Interestingly, after having published a critique of global environmental management, one of the authors found his one-man-NGO listed as a member of the 'Brundtland constituency'.

The second phase of the Center's activities began when it became clear, in September 1989, that the UNCED process was going to be launched and Maurice Strong was going to be UNCED Secretary-General. As a result, the Center redefined its mandate and priorities in order to feed into the UNCED process on the one hand and take advantage of the UNCED dynamic to promote itself on the other hand. Instead of making public relations for the Brundtland report, the objective of the Center now became the mobilization of the Brundtland constituency into the UNCED process. Lindner said: 'Maurice Strong was appointed to head UNCED and I went to him and said we would be happy to provide our assistance and support to mobilize in the broader constituencies'.[7]

Lindner must have been convincing. Indeed, in early 1990 UNDP offered Lindner US$10,000 to help sponsor an initial organizational meeting to further that purpose. As a result, the Center convened a strategy meeting for what it calls the 'independent sector', defined to include environmental and developmental NGOs, business and industry, trade unions, professional associations, scientific and academic institutions, women's organizations, youth groups, religious and spiritual groups, indigenous peoples' organizations, and other citizens' groups. The meeting took place in Nyon, Switzerland, in June 1990. The agenda was to mobilize for UNCED, in particular to prepare for the independent sector's participation in the imminent PrepCom meeting in Nairobi in August 1990.

In retrospect the Nyon meeting turns out to have been a key NGO gatekeeper meeting. Almost all the people who subsequently played a key role in Rio in the NGO and independent sector world were present there. But it would be just as correct to say that most of the NGO representatives that got invited to the Nyon meeting found it easy to get funding to attend the PrepComs and the Summit meetings.

In Nyon it was decided, among others things, to create a new body to coordinate NGO activities for UNCED, which would be structured, not surprisingly, according to constituencies and independent sectors. Also, there was a debate as to whether corporate industry should be part of the independent sector, which was finally accepted. An International Facilitating Committee – the IFC – was thus created as a coalition of independent sectors. The IFC is physically located within the Center for Our Common Future, but financially independent from it.

Members of the IFC are not individuals but organizations. Many of these member organizations, in turn, are federations or coalitions of other organizations. Moreover, in accordance with Maurice Strong's vision, the IFC was to organize all or at least as many independent sectors as possible, and feed them into the UNCED process. Members of the IFC include such organizations as the Brazilian NGO Forum, the Canadian Participatory Committee on UNCED, the Center for Our Common Future, CONGO, IUCN, EEB, indigenous people, the International Chamber of Commerce, Cable Network News (CNN), the World Conference on Religion and Peace, the International Committee of Scientific Unions, the Green Belt Movement, the International Confederation of Free Trade Unions, ANGOC, representatives of women's organizations, etc. As one can see from this list, environmental representatives were actually a minority. Among them, many defended a New Age vision of ecology, for example, the US Citizens' Network, or a global environmental management vision, e.g. the International Council of Scientific Unions (ICSU) and IUCN. Few actually represented a political ecology vision as, for example, did ELCI.

The IFC's stated aim was to assist organizations and networks in the independent sector to define their roles in UNCED, to promote fair and effective participation in UNCED on the part of the independent sectors, and to provide a forum for dialogue among the independent sectors. In more concrete terms, the IFC facilitated NGO access to UNCED, organized information briefings for NGOs, and organized the '92 Global Forum in Rio.

Overall, the IFC is certainly not a success story: important environmental and developmental NGOs refused to play along with the terms outlined by the IFC, the IFC itself became bogged down in procedural questions, and in Rio there were even some questions raised about its financial management. As a result, the IFC increasingly came under attack from the environmental and developmental NGO community, especially the ELCI and other Third World NGOs, which criticized, in particular, the IFC's bias towards business. The fact that the Center for Our Common Future practically took the IFC over by appointing Warren Lindner as its chairperson was also heavily criticized.

FROM THE PARIS NGO MEETING TO THE INTERNATIONAL NGO FORUM

At the above-mentioned Nyon meeting it was decided, among other things, that the IFC was to organize 'a pre-UNCED meeting of all interested groupings to concentrate on the official conference agenda to be held approximately six months before the Conference.' However, 'the ELCI board members felt that the ELCI, and not the IFC, should organize such a pre-Brazil meeting and that the IFC should play only a facilitating role'.[8] The ELCI's opposition is quite understandable since the ELCI had organized regional meetings on people's participation in environmentally sustainable development long before the Center for Our Common Future. As a result, the ELCI left the IFC and started to define its own strategy for facilitating the participation of environmental and developmental NGOs as well as community groups in the UNCED process.

An international Steering Committee was set up to guide the ELCI's work, meeting for the first time in August 1990. This International Steering Committee was co-chaired by the ELCI, the Brazilian NGO Forum, and Friends of the Earth. Overall, this Steering Committee gathered mainly those NGOs that opposed the IFC, most of which were political ecology groups from the North and participatory development NGOs from the South. A second meeting was held in Cairo in November of 1990. By then the participants included representatives from more than forty NGOs from thirty countries. The meeting discussed, in particular, the document that the Steering Committee intended to elaborate. The Committee considered that this document should focus on 'identifying local solutions to global problems which can contribute to changes in lifestyle, consumption patterns, etc.'[9] The Steering Committee's perspective was much more coherent than that of the IFC, but it was also more ideological, confrontational and political, focusing basically on grassroots and people's initiatives against the governments.

Also, at the Cairo meeting it was decided to take up the French government's offer to sponsor a Global NGO Conference in Paris in December 1991, preparing for the Rio conference but also serving as a platform for discussing the above mentioned document in a large NGO gathering. The French offer was contingent upon the Steering Committee's selecting 850 participants, which created all kinds of conflicts among NGOs. The so-called Paris NGO Conference took place just before Christmas 1991. It led to an (unpublished) NGO position paper entitled 'Roots for Our Future', containing a synthesis of NGO positions and a plan of action dealing in particular with climate change, biodiversity, forestry, biotechnology, GATT, resource transfer, institutions, and lifestyle from a grassroots perspective.

After the Paris NGO meeting the original Steering Committee was dissolved, and it was unclear where this alternative process was leading to. Some wanted to have an NGO/social movement gathering to be held in Rio prior to the UNCED Conference. Others wanted an International Civil Society Conference to be clearly separate from and different to the Global Forum. Finally, an International Coordinating Group was set up. Overall, its members were still more grassroots oriented than were the members of the International Steering Committee. They included representatives from ENDA (Senegal), APPEN (Malaysia), the Green Forum (Philippines), the Latin American Ecological Pact (Chile), the Youth Consultation for UNCED (Costa Rica), the US Citizens' Network, the Canadian Participatory Committee for UNCED, and the Polish Ecological Club, as well as two representatives of the indigenous people, from ELCI, the International Council for Voluntary Agencies (ICVA), the International Youth and Student Movement of the United Nations (ISMUN), the Third World Network, and the Brazilian NGO Forum. As a result of this whole alternative process, an International NGO Forum (INGOF) was held in Rio separately, yet within the Global Forum.

COOPTATION

No matter what all these NGOs and alternative organizations ultimately did in terms of content, they all conformed to what the UNCED establishment wanted them to do, i.e. mobilize for UNCED. If the IFC and the Paris process were certainly the two most massive efforts to mobilize for Rio, there were many parallel efforts on regional and national levels, as well as in other sectors. Whether it was the US Citizens' Network on UNCED, the Brazilian NGO Forum, the Canadian Participatory Committee for UNCED, in many others, the idea was always the same: getting mobilized and organized so that the respective constituency's voice would get heard by the governments and their negotiators.

The IFC certainly was the most favoured partner of Maurice Strong and the UNCED secretariat. However, after the incident with ELCI and the setting up of a parallel NGO process, it became clear that the IFC on its own could not effectively mobilize all independent sectors. As a result, other sectors and other groups started to organize themselves. The scientific sector, for example, worked under the leadership of ICSU and with the support of the Norwegian Research Council on Science and the Humanities and the Third World Academy after the May 1990 Bergen conference towards a Global Science Summit. The Summit was finally called ASCEND 21 – 'Agenda of Science for Environment and Development into the 21st Century' – and held in Vienna in November 1991. But most successful in mobilizing for and influencing UNCED was certainly the business sector, which is discussed in Part III.

Indeed, probably the most impressive phenomenon in and around Rio was the fact that everyone wanted to be part of it. Whether this was deliberate or not, the fact is that Maurice Strong managed to create, and the UNCED secretariat to implement, a structure through which organizations from various sectors were almost forced into mobilizing themselves, trying desperately to make an input into the Rio process and the corresponding documents. Beyond seeking to create a dialogue, this structure basically replicated the US model of 'democracy' by which constituencies mobilize and organize in order to lobby the establishment.

The more money one spends and the more professional this lobbying effort is, the more likely it is that the group will make an impact. In the next chapter we examine whether and how this lobbying effort paid off, and for whom it actually did.

At this point, however, it is important to highlight that the primary outcome of this mobilization exercise for UNCED was not, by a long way, the UNCED documents. The primary outcome has been the increased legitimation of governments, and spotlight visibility for UNCED and perhaps for Maurice Strong himself.

Indeed, whether they promoted the Brundtland report's view and sought to mobilize citizens or NGOs into UNCED, or whether they were trying to do exactly the opposite, all NGOs and other agents involved in and around UNCED became caught in what could be called the 'UNCED visibility trap': no matter whether they sought to promote or protest against the idea of sustainable development, whether they sought to feed into Rio or organize alternative meetings, they all did what Maurice Strong and before him probably Gro Harlem Brundtland had wished for, namely increased the visibility of the UNCED process. Moreover, many of the NGO coalitions themselves had a stake in this, as their own visibility had come to depend on UNCED. In this sense, the whole NGO mobilization process can be seen as a means to use NGOs for public relations purposes. Most of them gladly participated.

But there were certainly other purposes as well. To recall, in the framework as promoted by the Brundtland Commission and implemented by and via the UNCED process, global environmental problems will ultimately be solved, it is said, once the governments of the world have established a dialogue among themselves as well as with the main non-governmental actors. It was therefore essential to have the appropriate, i.e. the most influential, dialogue partners associated with the UNCED process. In general, these are the ones that can speak on behalf of a powerful constituency. In this respect, the concept of 'independent sector' is very typical: the concept carries a technocratic bias, as it makes the assumption that the world leaders – i.e. the heads of governments – express the public interest, whereas the independent sectors aggregate private interests. This concept implicitly states also that all interests are by definition private, and they therefore all have an equal right to be heard by the world's leaders – provided, of course, they represent a powerful constituency and they have the means to make themselves heard.

Not surprisingly, some partners in dialogue are more privileged than others in becoming associates in global management. The IFC as well as the UNCED secretariat promoted some independent sectors' organizations as privileged working partners of the UNCED process, whereas others, less organized and/or less powerful agents, were actually screened out. Not surprisingly, it is the business sector that has profited most from this model of American democracy. As a matter of fact, business and industry started preparing themselves for UNCED in 1984 – the year of the first World Industry Conference on Environmental Management. They ended up in 1990 with the creation of a Business Council for Sustainable Development. Its chairperson, the Swiss billionaire Stephan Schmidheiny, was appointed by Maurice Strong as his personal adviser. It comes as no surprise that the new global politics, stressing interpersonal dialogue and minimizing the role of change in socio-economic structures, promote the best organized and financially most potent 'independent' sector as UNCED's privileged working partners. In addition to business and industry, this has also been the case as regards some Northern establishment oriented environmental NGOs, in particular IUCN, WWF, and WRI.

Meanwhile, other organizations and NGOs got bogged down in their internal mobilization and organization processes. This was particularly true of environmental and developmental NGOs which are, by definition, much more heterogeneous and have, as we saw in the previous chapter, ideologically different trends. This model of UNCED mobilization weakened them rather than strengthened them. Compared with others, in particular with the business sector, environmental and developmental NGOs were much worse off after UNCED than before. So what, exactly, did NGOs achieve in UNCED?

6

WHAT DID ENVIRONMENTAL
NGOs ACHIEVE?

Let us pretend that you, the reader, are an Indian activist with a strong knowledge of forestry issues. You have been invited to attend the Fourth PrepCom in New York in late March 1992 by a sympathetic American forestry group with which you have exchanged information for several years and which has sent visitors to your community forestry project in India. Your American friends believe that you may be able to provide valuable input into the process. You inform an old friend in England who has written to you to say that she had attended the Nairobi meeting and will be going to New York too, and you agree to meet up.

Coming to the USA presents a problem. The US embassy will not issue you a tourist visa, nor will the Indian government allow you to take more than US$500 out of the country. Given that you expect to spend six weeks away, you have heard that this will not be enough money. Even this US$500 is a large sum of money because it equals half a year's wages for you.

You write to your American group which writes a letter of invitation for you and guarantees to support you, and then you go back to the US embassy. The visa is still not automatic, so you return for an interview and provide evidence that you have a job to return to. All this has taken up almost a week of your time.

Finally, you leave for the USA and make a stop in London where your English friend meets you. Since she has never been to New York either, she has checked with the embassy which says she won't need a pre-approved visa as she can obtain it upon arrival, from the US customs. Finally you arrive in New York. It takes you almost twice as long as other travellers to get through immigration because they are suspicious of your halting English. Early next morning the two of you go to the NGO centre at 777 UN Plaza where you are to meet the American forestry group. Then you are quickly shepherded into the UN.

Registration takes a matter of minutes because the American group has thoughtfully pre-registered you and then you head down into the maze of UN corridors to the first meeting of the day, a caucus of all the NGOs attending the meetings. Outside the meeting room you encounter a Costa Rican NGO representative who stops to ask you for something, but not knowing any Spanish, you shake your head helplessly. You go into the packed room where over 200 people are waiting for the proceedings to begin, and you notice that the Costa Rican has followed.

Had you been at the US Citizens' Network reception a few nights before, you would have recognized many of the faces here today. But you weren't there and this morning you wonder at the fact that all the people here seem to know each other quite well. You catch a few words here and there as people whisper about previous meetings in Geneva and Nairobi and you feel quite an alien. But the meeting is about to begin. Somebody starts by listing all the different meetings that were held the day before and you begin to wonder if you will ever understand what is going on; until you notice the Costa Rican beside you, who is plainly lost.

After the hour-long meeting, everybody speeds off to their little meetings with government representatives or to attend the sessions. You feel quite lost in this underground maze until you catch sight of your English friend again. She has discovered that, unlike in Nairobi, there is a formal NGO group that meets daily to examine the latest governmental proposals for a forest treaty and prepare a critique of it. Some representatives then take these ideas back to their governments to get them to present their ideas to the other governments. She has also discovered that it is possible to attend the informal governmental discussions. You elect to go to the governmental meeting and agree to meet for lunch and exchange observations.

Off to conference room 4 where the informal meeting is taking place. Unfortunately the guard does not let you in because you don't have a special ticket to attend the meeting. This takes half an hour to get and involves getting lost at least twice in the building. Finally, you get into the meeting to discover that there are only two other NGO representatives in a gallery that can seat 200. You wonder what all the fuss is about, particularly when you return the next day and the next day to discover that nobody else attends these meetings.

Somebody gives you a spare copy of the forestry document under discussion – curiously named A/CONF.151/PC/WG.1/CRP.14/Rev2 – and you settle in to listen. At the side of your seat you discover a little knob that allows you to listen to the discussion in the original language or in any one of the six official languages of the UN: English, French, Spanish, Russian, Chinese, and Arabic. At first it seems a little strange: the delegates are not arguing about forests, but about brackets. Eventually you discover that the document you are reading was prepared by an expert and the 150 delegates are editing it into a document that they can all agree on. The brackets are put around the parts that they cannot agree upon. There is general laughter as the meeting chairman, Charles Liburd of Guyana, says: 'I understand that 150 copies of the bracket-ed document have been circulated'.

Frustrated, you wander off to another meeting. This one is about oceans and you listen in. The Indian delegate raises his hand and you listen with interest: 'As I have said Mr Chairman, ad nauseam, the phrase "where appropriate" should be added to "support from international groups"'. Another NGO representative who has been following the debate explains to you that this is to ensure that Northern activists cannot interfere with the sovereign right of Southern countries to choose their own development plans.

Thus might have been your first day at the New York meetings. Many other similar days might follow. The fictional account above is a composite of true stories of people the authors encountered at the meetings, as well as of their own experiences. Incidentally, both quotes from delegates at the meetings are real ones. Moreover, not only is this story of our Indian activist real, it is also highly symptomatic. First it shows, as we highlight in the next section, that in certain ways NGOs have quite easy access to the negotiations. But it also shows, as we detail in the subsequent section of this chapter, that NGOs hardly make a difference.

NGO ACCESS TO UNCED

Above we have argued that NGO access to the UNCED process was deliberately made easy, and sometimes even paid for by UN agencies or other donors. We think that this easy access was ultimately detrimental to the NGOs and confused them. In this section we would like briefly to discuss the major areas where NGOs did have access to the UNCED process, namely access to the negotiations, as well as becoming part of government delegations.

Though access to UNCED for NGOs was quite easy it was also highly confusing. Many NGOs did not understand the lobbying process that would have given them access to the negotiations. Others did not speak English. Still others were simply overwhelmed by the complexity. In order to influence the negotiations NGOs had basically three possibilities: to speak up in the sessions where this was possible, to submit written statements to the negotiations, and to establish personal contacts with the delegates.

Briefly, this is how lobbying was actually conducted. At the four PrepComs NGOs sat down every morning to plan strategy and to brief each other about what they had heard the previous day. Not everybody attended. During the day some met privately with government or UN officials, and attended the governmental discussions. When they got permission, they made statements about the subject under debate. In each of the fora they tried to make suggestions for textual changes to the agreements being discussed.

Separately they met in small groups to learn more about particular issues. Governments were invited to attend and face their questions. In Nairobi there were almost no such meetings, but by the third PrepCom in Geneva there were already two to three meetings happening simultaneously at most times of the day. Finally, in New York, there were up to six meetings happening at the same time and the UN agreed to allow the NGOs to take over its meeting rooms after the delegates had gone home.

NGOs did not, however, get into all the government meetings. As they progressed, the governments set up special meetings called 'informal-informals' where admission was restricted strictly to delegates and NGOs which were on government delegations. But all of this was confusing and frustrating to many of the NGOs which, being new to the process, understood little of the lobbying.

In a survey that Ann Doherty from the International Institute for Applied Systems Analysis conducted right after the Earth Summit among the NGOs, three-quarters said that they were not satisfied with the access granted to them.[1] Only 12 per cent of the NGOs that responded were satisfied with the amount of speaking time allotted to them. Yet it is quite possible that many respondents confused access and influence. Indeed, another measure of access indicates that NGOs were well organized and functioning within the system. The average quantity of written interventions per NGO hovered between five and six in a broad range from zero to fifty. Of course, the quality of these interventions is undetermined and it remains unclear what the delegates actually did with these interventions. But this nevertheless shows that NGOs did have the possibility of access to the negotiations.

Overall, one can say that the speaking time was either inadequate or not well used, whereas lobbying through meeting with delegations and submitting written statements was considered more effective by the NGOs. But most effective by far, according to the NGOs, was direct personal contact in order to create a good relationship as early in the process as possible, preferably beginning in the home country.

But those who said that their views were 'often' incorporated into the documents were few, and tended to have been either on delegations or had some other special status in relation to delegations. Overall, being successful, i.e. having some influence on the negotiations, was basically a matter of good relations with government delegates and the secretariat. Environmental NGOs with such good relations were the WWF, IUCN, the WRI, and the Big 10. The Third World Network had such good relations with the Malaysian delegation.

But even then the influence of NGOs on the final wording of the UNCED documents was minimal. Says one NGO representative in the aforementioned survey: 'They at best took some formulations, but never the intentions.' And in the opinion of an editorial writer on *Crosscurrents*, the NGO newspaper, 'they used fragments of the text without the spirit of the whole recommendations'.[2] And even Mark Valentine, issues director of the US Citizens' Network, arguably one of the most powerful Northern lobbying groups during the Summit process, admitted:

Most NGOs would have to concur that citizens' groups barely scratched the surface of the official documents. Bits and pieces were tinkered with and modified here and there, but the structure of the agreements, the context within which they were considered, and the level of political and financial investment, all conformed to governments' expectations, not NGOs.[3]

Being on a government delegation, therefore, was a more direct means to influence the UNCED process and documents. As UNCED went on, more and more countries appointed representatives of what they called the independent sector to their national delegations. But they could basically represent any sector varying from business to academia to environment and development NGOs. Canada was apparently the first country to put NGO representatives on its national delegation. This occurred during PrepCom I in Nairobi, where Canada was the only country doing this. Moreover, Canada set yet another precedent by letting the NGO representative speak in a plenary session. By PrepCom II at least eight countries, almost all from the North, had appointed NGO representatives to their delegations. They were Australia, Canada, Norway, the Netherlands, the United Kingdom, the United States, the Commonwealth of Independent States, and India. And in Rio about fifteen governments allowed NGOs to join their delegations as observers, attend morning briefing sessions, and even join them at the negotiating tables at the government discussions, where they could make minute-by-minute suggestions about the documents being discussed.[4]

Nevertheless, there seem to be considerable differences from one country to another as to the exact role NGO representatives played on national delegations, as well as to the degree they were integrated in the delegation. Canada, for example, asked its NGOs to provide advice and expertise, whereas others, like France, asked NGOs to represent the government. There was also a difference in terms of the information governments provided to their NGOs. Dawkins reports that during PrepCom II, for example, 'the British delivered three inches of official briefing papers to their NGO delegates, whereas the US provided no instruction whatsoever. Most of the NGO delegates had been given more specific instruction from their constituent organizations than from their respective government'.[5]

Overall, environmental NGO representatives on government delegations again basically represented mainstream NGOs. But as already mentioned above, even these NGOs do not seem to have made a major difference in terms of the final wording of the UNCED documents.

DID ENVIRONMENTAL NGOs MAKE A DIFFERENCE?

As we have seen, NGOs had considerable access to the UNCED process. They also had substantially mobilized in order to be able to take part in this process. But did they actually make a difference within or outside the UNCED process?

We have tried to show that the UNCED documents were hardly affected by the various NGOs. This is with the exception of some mainstream environmental NGOs, whose positions were so close to the positions of the governments that their distinctive impact can hardly be detected in the texts.

The governments, however, did offer NGOs a specific chapter in the mammoth Agenda 21. The chapter discusses the creation of a 'real social partnership' between governments and NGOs and says there is a need to provide mechanisms for the substantial involvement of NGOs at all levels from policy to decision-making to implementation. If one thinks of the fact that at PrepCom I NGOs were not even mentioned, the mere existence of such a chapter is already in itself a major achievement. However, this chapter, like the entire Agenda 21, is unlikely ever to be used unless NGOs can persuade governments to implement it in their home countries.

But if NGOs made no difference within the UNCED process, did they at least make a difference outside it? Three aspects must be looked at in this respect, namely the Global Forum, the Alternative Treaty writing process, and the contacts among NGOs.

The Global Forum, in which about 30,000 people from all over the world participated, is probably best described as a circus or a colossal mess. All kinds of activities went on during the Global Forum, from theatre and dance to commercial events, New Age celebrations, and celebrity appearances; from sectorial alternative negotiations to a protest against the World Bank; from exhibits to a backwards march to the Rio Centro to symbolize NGOs' opinions of the progress made in Rio. The excitement was heightened by the near-bankruptcy of the event: with a debt totalling US$2 million, the electricity, translation, and meeting areas were saved at the last minute only by an emotional fundraising drive. Many NGOs were enthusiastic about opportunities to meet other like-minded people; others were disgusted at the frivolous tone of the event.

The main grouping within the Global Forum was the International NGO Forum (INGOF), also known as the International Forum of NGOs and Social Movements. It was a grouping of progressive and political NGOs, whose origin can be traced back to the Paris NGO meeting in December 1991, which was jointly sponsored by ELCI, Friends of the Earth, and the Brazilian NGO Forum. It gathered together political ecologists from the North and grassroots development NGOs from the South. The main focus of INGOF was on drawing up over thirty NGO 'treaties' on subjects ranging from climate change and biodiversity to poverty and racism.

Opinions on the thirty treaties were not very favourable, apart from those who wrote them. Some NGOs called the process 'dogmatic', 'stifling creativity', 'lopsided', and failing in crucial linkages between treaties. One NGO commented: 'The negotiations were done by people who could afford to pay a ticket to Rio. Can we hold these treaties out for the world to see and say they represent the small and large NGOs around the world?'.[6]

But the main problem with the treaties, however, was that no one seemed to know what exactly they would 'do' with them other than 'use them in the post-Rio work'. Though there are some substantial differences between these alternative treaties and the official ones, it is unlikely that they will ever have any practical effect.

Therefore, the main success of all the NGOs' parallel activities in and around the UNCED process might be limited to contacts and mutual learning. In the opinion of Martin Khor, president of the Third World Network, the success has been the evolution of NGO opinion, particularly in the North. He said:

The UNCED process forged new and stronger links between Northern and Southern groups, between development and environmental activists. It would now be difficult for environmentalists to stick to wildlife issues or population, without simultaneously addressing international equity and global power structures. A major step forward has been the increasing involvement of Northern based environment groups like Greenpeace, WWF, and Friends of the Earth in economic issues such as terms of trade, debt, and aid . . .[7]

Finally, the image NGOs gave in and around Rio was either absent, confusing, or negative. Indeed, the average person on the street following UNCED through the media had no idea that NGOs were lobbying at the Summit or that they could have been part of this process. If they did get any air time, their analytical back-up was largely confused. For example, because George Bush was everyone's undisputed bad guy, most people took the position that what Bush was against, they were for. Demonstrations were held and press releases demanded Bush's signature on the biodiversity convention. Yet the very few NGO analysts who had been following the complex negotiations on the convention were themselves calling for countries not to sign it. Similarly, NGOs lambasted Northern countries for not giving more aid, while criticizing past aid, claiming that environmental problems would not be addressed merely by giving money and fiercely attacking the main institutions through which the new loans and grants would be given.

There was also a lack of a clear message on issues of debt and trade. NGOs denounced Third World debt as an instrument of Northern imperialism, while demanding more loans for the South. They also called for the South to receive more money for the export of its commodities, while denouncing export-oriented development strategies. If these contradictory positions had been presented by different NGOs it would have been understandable, but often these arguments were being presented by the same organizations.

Not surprisingly, Greenpeace made by far the best use of the media of any NGO. Wide coverage was received, especially in Brazil, for the Rainbow Warrior's blockade of a paper mill and a nuclear plant, as well as for the hanging of a huge banner on the side of Sugarloaf Mountain at the end of the Summit with a picture of the Earth and the simple message 'Sold'.

However, the overall image NGOs gave at Rio was negative. The *Financial Times'* final summary, for example, included NGOs on its list of losers at the Summit. They were, it reported, 'shut out by the politicians, and spent most of their time at their Global Forum 50 km away, where they ran out of money and had their electricity cut off'.[8] Other summaries reflected similar images. The *New Scientist*'s summary said that NGOs 'appeared marginalized, their lobbyists wandering round in ever increasing gloom. The greens had their stunts and photo opportunities but little more'.[9]

Why these negative images? Where did the journalists get their ideas from? While the media must be blamed for ignoring the NGOs and not following the issues in detail – not one newspaper, television or radio station sent a reporter to cover all the PrepComs – the NGOs must also take blame for focusing so much on lobbying on the inside, where no one could see them, instead of being a voice for the millions they were supposed to be representing. And if they did not have the thousands of voices to make their presence felt, why did they choose to sit down and compromise themselves into oblivion instead of taking on the media?

This, in our view, was the result of a long-term transformation of the Green movement worldwide, combined with the very way the UNCED process was set up, as a means of reducing potential protest by feeding people into the Green machine. As a result, NGOs and the movement fudged what should have been their finest hour.

THE END OF PROTEST?

Maurice Strong, who had already orchestrated the UN Conference on the Human Environment in 1972, took at least one lesson away with him from Stockholm: avoid protest and confrontation. As part of the political context of the late 1960s and early 1970s, including the ongoing Vietnam War, the Stockholm Conference was marked by heavy protest. At least part of this protest can be explained by the fact that civil society was basically shut out from the Stockholm process. Overall, the Stockholm Conference was characterized by heavy confrontation between activists of all sorts and governments. This was not going to happen in Rio. And indeed the overall climate was one of consensus and cooperation.

With the exception of one demonstration in Rio de Janeiro which brought together 50,000 people in downtown streets, most protests drew a few dozen people. With the exception of the treaty-making process, which attracted 2,000 NGO representatives, most of their meetings attracted only a few dozen. Rio, after all, was a highly individualistic event, reflecting the overall New Age spirit.

And despite the fact that the media covered some protests – especially when the UN security guards dragged forty youth activists from Rio Centro and detained ten, when these images were flashed on to ABC, CNN, Australian, German, and Hong Kong TV, with pictures in the *Los Angeles Times*, the *New York Times*, the *San Francisco Examiner*, the *San Francisco Chronicle*, the *Washington Post*, and the *Village Voice* in the USA, *The Independent* and the *Guardian* in England, and *Libération* in France – their actions were almost certainly forgotten by the public at large within days of the occurrence.

To be sure, access to the heads of government at the Summit had been carefully limited to those who had registered in advance, so large protests were almost impossible. Then, of course, there is the fact that the Brazilians had pretty well sealed off the conference centre with 35,000 troops, tanks, and helicopters for four miles in every direction.

Finally, the UNCED process had been set up from the very beginning in a way that made people feel they were part of it, a game most NGOs gladly and very actively played. And the few groups that had actually criticized or even opposed the Rio process from the start never showed up and did not bother to participate.

On some occasions, UN agencies, as well as national, governmental, and private donors, even paid money so that people would become active participants in the UNCED process. This was the case with UNDP, for example, which spent approximately US$682,000 on sponsoring NGOs, US$475,761 on three programmes in 1990 and 1991, and then US$206,400 in the final six months up to and including Rio. The money for this had been raised largely from the governments of Norway and the Netherlands.

The three main projects that UNDP paid for were to support NGOs to go to the meetings and it gave US$10,000 in assistance to NGOs in twenty-three Southern countries to enable them to generate interest in the Summit at home. There was also support for a special meeting on poverty and the environment in Geneva in March 1991.

From this UNDP funding also sprang two major drives among the Southern country NGOs. First, it strengthened the Third World Network – an existing umbrella organization of Southern country NGOs who were already working on the issues. Second, it created a demand for a special emphasis on issues of poverty, which then spawned a protest against the World Bank and demands for alternatives to it, notably under the aegis of Maximo Kalaw, President of the Green Forum of the Philippines.

Third World Network brought some heavy hitters to the various PrepCom meetings – Vandana Shiva, an eco-feminist from India, Martin Khor, president of TWN at its headquarters in Penang, Malaysia, Chee Yoke Ling, the head of Friends of the Earth in Malaysia, Charles Abugre, an economist from Ghana, and Daniel Querol, a biologist from Peru. All of these were recognized experts in their fields and they churned out a series of briefing papers to counter government ideas.

Most of their criticisms were directed against the World Bank, the IMF, GATT, and of course the USA. They were silent about UNDP. UNDP arranged for them to confront the Bank at its meetings in Washington, DC, when it agreed to meet NGOs at the new GEF participants' meeting (unfortunately, because of a prearranged South strategy meeting neither Vandana Shiva nor Martin Khor could attend). UNDP even sat down in private with TWN and briefed them on the key issues that the World Bank could be swayed on. Meanwhile Maximo Kalaw separately led a group of NGOs to put together a common position paper on poverty and call for a new institution to be set up, called the People's Bank.

What is wrong with this? Of course, UNDP and others must be lauded for building up Southern capacity. But at the same time NGOs were being tricked into giving some support to an institutional process that had created the problems they were raising to begin with. By stressing (and financing) their criticism of inequity and poverty, UNDP was using these and many other NGOs to build up a South–North conflict, whose only solution, of course, turned out to be more development. By letting themselves be mobilized along these lines, many Southern NGOs and in particular the TWN directly played into the hands of the development establishment internationally and even more so nationally.

And this is actually quite symptomatic of the overall outcome of the UNCED process: the mobilization of peoples and NGOs to participate actively in the UNCED process, while not letting them influence the outcome, has led to an overall legitimation of a process that is ultimately destructive of the very forces that were mobilized. Some Southern NGOs and NGO representatives through their participation in this process quite logically became coopted. This added some well needed fresh blood to the old development élite, which had already absorbed the mainstream Northern NGOs such as WRI, WWF, IUCN, and the Big 10.

If there was no substantive outcome in terms of conventions and documents, UNCED was at least an exercise in mobilization and cooptation, weakening the Green movement on the one hand while identifying and promoting potential opponents – mainly from the South – on the other. This UNCED has done successfully by extending the US model of 'democracy' to the planetary level. This model is basically a lobbying model to which theoretically everybody has access, yet only the strongest ones are successful. As already happens within the USA, this model has a high potential for mobilization – especially by the media – while simultaneously promoting the financially most powerful. Many people and NGOs, indeed, did get mobilized without being able to lobby at all. Others, as in the case of the Southern NGOs, were mobilized either by the UN itself or by other sponsors, such as the big Northern NGOs, which raised money from their governments and foundations. And of course the largest number of people at the Summit were professional lobbyists from Greenpeace, Friends of the Earth, the Big 10, and their European equivalents. Greenpeace, for example, had thirty professionals at the New York PrepCom (though not all the time), more than all except a half a dozen of the governments.

On the other hand, very few real grassroots or community groups went to Rio or joined the two-year process. This is because becoming part of the sort of lobbying system set up by the secretariat requires the activists to have a detailed knowledge of the UN and government bureaucracies and easy access to international telecommunications and travel. This effectively ruled out most community activists, especially in Southern countries. What is more, the lobbying system set the grassroots and community groups up against better funded corporate or special interest advocacy groups (including other big environmental NGOs), which do have access to all these facilities. No effort was made to ask communities what was wrong and ask them how to solve it. Rather, the most vocal NGOs were called upon and promoted to try to make them part of the top-down problem-solving process. As a result, they themselves became part of the problem.

It is, of course, unfair to blame these NGOs for the failure of the UNCED process. But it is legitimate to question their buying into the UNCED process without prior critical reflection. Anyone with a lucid mind should have seen that this system was set up for potential lobbyists who would follow the process from meeting to meeting, sit down and compromise, and legitimize it while doing so. As a result of the UNCED process, most environmental lobbyists have lost their innocence *vis-à-vis* their constituencies. Southern NGOs in particular, some of which are now quite alienated from their grassroots constituencies, have been driven into the arms and are now at the mercy of UN agencies, Northern governments, and, especially, their own governments. What is more, they can now, and almost certainly will, be played off against each other, thus ultimately weakening the position of the South.

In conclusion, let us offer an opinion we (the authors) formed as NGOs and governments were gathering in Rio – drawn from the original paper that went on to form the basis of this book:

The UNCED process has divided, coopted, and weakened the green movement. On the one hand UNCED brought every possible NGO into the system of lobbying governments, while on the other hand it quietly promoted business to take over the solutions. NGOs are now trapped in a farce by which they have lent support to governments in return for some small concessions on language, and thus legitimized the process of increased industrial development.[10]

Part III

BUSINESS AND INDUSTRY

The UNCED process has been a clear success from the perspective of business and industry, in particular big business, and more precisely transnational or multinational corporations. Business and industry are, in fact, the only sector that can claim success. Business and industry not only became entirely part of the UNCED process, they shaped the very way environment and development are being looked at. Indeed, in the absence of any intellectually coherent analysis of the present crisis and the solutions to it, the view of business and industry came to dominate. And since business and industry were actively promoted to be an integral part of UNCED, their view has rapidly spread worldwide.

This is totally different from what had happened in Stockholm twenty years before. According to Harris Gleckman at the UN Center on Transnational Corporations (UNCTC), at the 1972 Stockholm Conference on the Human Environment the role of the business sector was a single intervention lasting eight minutes by the International Chamber of Commerce (ICC). At Stockholm, business was, like the NGOs, basically left out of the process. Worse, it was an object of potential environmental regulations. Within the political context of the 1960s, business and industry were very clearly on the defensive. This was totally different in the UNCED process: Strong, who between Stockholm and Rio had himself become heavily involved in big business, made this sector become part of UNCED from its very inception. Business and industry were offered multiple opportunities to pay their way and have their say in the UNCED process. They had multiple occasions to lobby. Of course they took advantage of it and were quite efficient at it. As a result and unlike at Stockholm, business and industry were no longer objects of discussion, but 'partners in dialogue' to help solve environmental and developmental problems.

NGOs did not realize this at the beginning. However, they became increasingly worried about the fact that corporations were taking part in UNCED, but their activities were not being discussed at all. At various stages during the Summit's preparatory process, activists voiced their concern and by the time of the fourth PrepCom in New York, they suddenly realized that corporate pollution was not going to be discussed at all in the final documents in Rio. So Greenpeace joined hands with the US-based National Toxics Campaign and TWN, among many other NGOs, to condemn the environmental record of multinationals at a press conference, and the absence of this issue from the agenda.

When it came to light that business NGOs were also attending meetings and in fact lobbying behind the scenes, there was outrage on the part of the other NGOs. At the New York PrepCom, for example, there were bitter exchanges between the two groups at an NGO–government dialogue when a Canadian business lobbyist attended the meeting. Greenpeace's Summit coordinator Josh Karliner delivered an impassioned speech to one of the evening NGO meetings condemning industry in general.

Yet the bulk of industry was neither present nor lobbying at UNCED. The type of industry that got fed into, received visibility, and was promoted in the UNCED process was mainly multinationals or transnational corporations (TNCs). Many of them are heavy polluters and therefore particularly interested in the outcome of Rio. As we have shown earlier, the entire Rio process was set up as a lobbying exercise. And given the fact that TNCs were perfectly at ease with this lobbying model, which they had practised in many countries, it is not surprising that they turned out to be quite good at it.

Activists, of course, see corporate pollution as a major international problem and they wanted the Summit to make some firm commitments on regulating their activities. A commonly cited figure is that multinational corporations conduct 70 per cent of international trade and 80 per cent of foreign investment, and rival the military in terms of the pollution they emit and cause with their products.[1] The UNCTC, for example, notes that multinationals control 80 per cent of cultivated land for export crops worldwide and a mere twenty of them control 90 per cent of pesticide sales. They also control the major share of the world's technology and dominate key industries in the mining sector.[2] Moreover, they control the markets for most of the major products of Southern countries and are thus responsible for the unsustainable depletion of habitats and resources caused by the extraction and cultivation of these products.

Since TNCs are indeed the major agents in the global development and environment arena, the fact that they were made part of the UNCED process is justified. However, the way the process was set up and run made TNCs appear less and less to be part of the problem. As Rio came closer, they appeared more and more to have the solutions or to be the solutions to the kind of problems for which they were at least partly responsible.

In Chapter 7 we examine how TNCs got fed into and promoted by the UNCED process. We look at how business and industry prepared themselves for Rio, paid their way, and organized finally to take over the process altogether. In Chapter 8 we consider the ideological implications of this takeover. These are, in our view, much more worrying: we thus examine how business and industry have redefined environment and development issues to fit their needs and deeds, and we show why the business view of how to solve the global environmental and developmental crisis is fundamentally flawed.

7

PROMOTING BIG BUSINESS AT RIO

Business and industry have always had an inside track with governments. Over the years, business has become quite good at lobbying national governments, especially in the 'model' Western democracies such as the USA and Canada. More recently, lobbying by business and industry has also become a crucial part of EC politics in Brussels. In international negotiations such as GATT or FAO discussions on Codex Alimentarius, governments regularly take business executives along. To a certain extent this is quite natural and normal, since on many issues governments and industry pursue the same goal. In particular, they share the same core value, namely that industrial development is the foundation of modern society, and that it must be pursued at any price. If industry is perhaps focusing more on the production of wealth, government is focusing more on its distribution. But both are obsessed by economic growth. It is therefore only logical that many governments included business and industry advisers on their delegations to the PrepComs and the Rio conference.

Also, during the 1980s it became quite acceptable for environmental NGOs to solicit donations from private corporations and many of them make it a point to go to corporations and get money. By the end of the 1980s most Northern NGOs had levered substantial corporate contributions. Many of them even have joint programmes with corporations. The Environmental Defense Fund (EDF), one of the Big 10 environmental NGOs in the USA that follows the lobbying model of politics, which is predominant in the USA, for example, was praised by industry at Rio for its compromising stance. In recent years EDF has signed two major cooperation agreements with McDonald's and General Motors (GM).[1] Both of them agreed not to use the agreement for publicity and both allowed EDF publicly to criticize their policies. As a result, EDF suggested that McDonald's use recycled paper in their packaging. To GM it suggested that old cars be bought up by companies wanting to receive pollution credits for reducing pollution. These credits could then contribute to their required targets of reducing pollution under the US Clean Air Act of 1990.

While both suggestions are useful if they constitute part of an organizational learning process that will lead to much more profound and radical changes, we have good reasons to believe that these agreements basically pursue strategic purposes. While GM was getting all this free publicity from EDF, it was actually suing state environmental protection agencies in Connecticut, Massachusetts, and New York to limit new strict air quality laws. McDonald's did not need to go to EDF either to discover that recycled cardboard was more ecologically sound than Styrofoam. The problems of Styrofoam and advantages of recycling have been topics of common discussion all over the USA for years. But what did count for it was the implicit endorsement from EDF of its efforts. Moreover, it was of course not EDF that convinced McDonald's that it should use recycled cardboard, to stay with this example. The fact is that McDonald's policy change was a direct result of a campaign by the Citizens Clearinghouse for Hazardous Waste's Lois Gibbs, who had led the earlier, very successful, protest on corporate dumping in Love Canal, New York. Gibbs got thousands of schoolchildren to send their used McDonald's fast food containers back to the company.

Not only did EDF take undue credit for a change in McDonald's corporate policy which it did not have much to do with, but moreover they contributed to the fragmentation and erosion of the environmental movement. The point of these two examples is that by paying for NGOs that do not criticize them, corporations can marginalize the ones that do. What is more, they can get free mileage out of groups like EDF and portray themselves as compromisers and listeners, while their motivation remains strategic. EDF, of course, maintains that the agreement specifically gives it the right to criticize GM, but in fact it admitted that it was not doing that. Senior attorney Joe Goffman told us that EDF had different opinions on many subjects from GM, but that is hardly surprising. Moreover, anybody can criticize a company and that does not need a special agreement.

The precise problem is that this is not simply a matter of opinion. The question is whether EDF is ready and GM willing to engage in a process of mutual learning, the ultimate outcome of which should be phasing out from environmentally and societally unsustainable car production and car culture altogether. This example highlights very well, in our mind, the type of problems raised by the corporate sponsorship of NGOs. But the result is that the big environmental NGOs are becoming less critical of big corporate polluters, while the critical groups are being marginalized.

The next example brings us closer to what was going on in Rio as regards the role of business and industry in environmental matters. Right on time for Rio, the Swedish/Swiss multinational Asea Brown Boveri (ABB) launched its own journal *Tomorrow* dealing with global environment and development issues. On its board are representatives of major Northern and even Southern NGOs. This is symptomatic of the fact that up to and during the Rio Summit, business and industry, especially big business, were no longer influencing and lobbying the main agents. Rather, they were shaping the environment and development debate. Indeed, given the abdication of governments and the erosion of the environmental movement, the Rio conference became a platform from which business and industry, often with the help of public relations agencies, were offered an additional opportunity to shape the way the public should think about environment and development.

This is our argument: the new global reality of which UNCED is an expression and which it simultaneously helps to promote, is of a fundamentally different nature to the national realities on which governments had a monopoly, and where other agents such as NGOs and business could lobby governments. For most agents the global reality is something new: it is not at all clear yet whether national governments and most NGOs are up to the challenges of this new global reality. This is also true for the UN system which remains an organization of nation-states and which, moreover, has been set up with a different purpose to the one required by today's global environment and development crisis. Indeed, the UN was to guarantee world peace, which was to be achieved through accelerated industrial development. Today, war and peace have changed their very nature, and industrial development hits bio-geo-physical limitations. Therefore, even if the UN once was a coherent global agent, the philosophical basis of its actions has now eroded. The only currently functioning global agents are therefore TNCs.

UNCED set up a process through which TNCs were transformed from lobbyists at a national level to legitimate global agents, i.e. partners of governments. UNCED gave them a platform, from where they could frame the new global issues in their own terms. In this chapter, we show how the UNCED process was set up in a way that meant big business and industry would inevitably turn out to be the winners. We look at how business and industry systematically prepared themselves for the Earth Summit in anticipation of the future role they were going to play, and we also highlight how business and industry benefited from the way UNCED was run. Finally, we look critically at corporate sponsorship of the Earth Summit. This is not because we believe that corporate sponsorship played a decisive role. Rather, such sponsorship, in our view, illustrates the much more profound process of corporate takeover of the leadership in environment and development matters.

THE UNCED PROCESS FAVOURS POWERFUL LOBBYISTS

Although business and industry, in particular TNCs, took over the UNCED process, this was not a 'hostile takeover'. It was not, in our view, the result of a conspiracy, though public relations certainly helped. Rather, TNCs just did what they were supposed to, i.e. shape the outcomes of the UNCED process in a way that was advantageous for them in the long run. In doing so, they have been considerably helped by the way UNCED was set up to begin with, by Maurice Strong's active advocacy for business and industry, and by the absence of any other major global agent. Also, neither governments nor NGOs seemed to be willing or able to oppose this takeover. As a matter of fact, many essentially Northern governments were highly supportive of business and industry and offered themselves as a platform from which to lobby UNCED. The big Northern NGOs were, as we have seen above, already quite compromised with business and industry, and had already more or less agreed to the idea that business and industry should play a key role in solving the global environment and development crisis. The big Southern NGOs – essentially the Third World Network (TWN) – though highly critical of TNCs, played right into their hands by portraying the global crisis as a South–North issue, thus making everybody believe that this crisis was not the result of industrial development, but rather an issue of more equal distribution. All other NGOs, finally, were scattered and fragmented, the result of both the set-up of the